Water Resource Management

Water Resource Management
Institutions and Irrigation
Development in India

A. VAIDYANATHAN

OXFORD
UNIVERSITY PRESS

OXFORD
UNIVERSITY PRESS

Oxford University Press is a department of the University of Oxford.
It furthers the University's objective of excellence in research, scholarship,
and education by publishing worldwide. Oxford is a registered trademark of
Oxford University Press in the UK and in certain other countries

Published in India by
Oxford University Press
2/11 Ground Floor, Ansari Road, Daryaganj, New Delhi-110002, India

First published by Oxford University press 1999
Oxford India paperbacks 2001

ISBN-13: 978-0-19-565884-2
ISBN-10: 0-19-565884-1

Printed in India by Repro Knowledgecast Limited, Thane

For K.N. Raj

PREFACE

The expansion and improvement of irrigation facilities has been a central feature of India's agricultural development strategy for nearly five decades. Almost throughout the country the bulk of rainfall is concentrated in the course of three to four months; even during this period, over large parts of the country, there is either too little or too much rain to sustain healthy crop growth: everywhere, the timing and quantum of rainfall is variable and unpredictable. And outside of the monsoon season, the rainfall and such of the monsoon precipitation as is retained in the soil, is inadequate to sustain any crops.

The function of irrigation is to protect crops from unpredictable variations of rainfall in the monsoon season and to harness the surplus of this season to augment the quantum and duration of moisture available for agriculture irrespective of the extent of local rainfall. In doing so, it helps stabilize the yield of traditional crops in the monsoon season; enables relatively low rainfall areas to cultivate better yielding crops requiring more water and better assured water supply; and permits crops to be grown even during the non-monsoon season. All this contributes to higher production from a given extent of cultivated land.

The quantum of nutrients that plants can absorb, and the efficacy with which they are converted into biomass, is also crucially dependent on soil moisture conditions: careful management of the soil moisture regime is specially relevant for realizing the full potential of the new hybrids and high yielding varieties of plants that are more sensitive to moisture stress than traditional varieties.

Given these considerations, it is not surprising that Indian Plans attach so much importance to irrigation. Massive investments have been made in this sector since Independence. The performance has however belied expectations both in terms of the pace of development and use of facilities, and of their impact on productivity of land. At the same time, the negative consequences (in terms of displacement of people, submergence of forests, waterlogging and salination of agricultural land) have been greatly underestimated. New problems have arisen from the rapidly growing demand for water for non-agricultural uses (the resulting competition with irrigation becoming quite acute in some regions and increasing everywhere), from the increasingly widespread pollution of water resources by industrial effluents and sewage; and the progressive and widespread lowering of groundwater table.

Many of these problems are recognized in official documents (such as five year plans, publications of the Central Water Commission, and reports of various committees). They are the subject of extensive studies by scholars and experts outside the government. The debate generated by activists opposing the current approaches to water resource development form the third important corpus of the critique on the way water resources are currently managed, but these issues tend to be debated piecemeal. There is too much polemic and too little in terms of reliable facts and insightful analysis. The technical complexity of the problem tends to be underrated.

Suggested solutions tend to oversimplify by focusing on particular aspects. Government engineers stress techno-managerial measures, while some environmentalists see the solution in small-scale community managed water conservancy works. Of late, privatization of water rights, turning over systems to beneficiaries, and use of the market mechanism to achieve efficient allocation are being advocated.

This collection of essays seeks to provide a more comprehensive, holistic view of the problem of water resource management, highlighting its technical complexity, the necessity for an integrated perspective in harnessing the various sources, and matching them with various uses; the importance of devising appropriate institutional arrangements in combination with better, more imaginative use of technology. It is my contention that institutional deficiencies are at the root of many of the problems, and that restructuring the organizational and management structure is an essential precondition to tackling them. It is therefore important to focus on weaknesses of the existing arrangements and the directions in which they need to be restructured to facilitate more efficient and sustainable use of water.

The appropriate institutional structure, however, depends on numerous interrelated factors: agro-climatic conditions, agrarian organization, the social structure of user communities, the technology of agriculture and water management and, of course, the social goals of state policy. Since all these vary greatly over space and time, one cannot think of any standard, uniform institutional structure. A comparative study of structure and changes in irrigation—and more generally water control—institutions in different contexts could be valuable in devising effective reform strategies appropriate to varying conditions.

This being the aim, the first essay in this collection attempts to provide an overview of water control institutions from this perspective. It highlights the importance of agro-climatic conditions in determining the need for, and forms of, irrigation; how the latter shapes the way irrigation organizations are structured; and the fact that the evolution of irrigation, and the way it is managed, are determined by the interactions between changes in the technology of harnessing and conveying water, the technology of agriculture (which affects the returns to water), relative costs of alternative ways of meeting demand, the characteristics of agrarian structure and the wider socio -political environment. Experiences of different Asian countries are used to illustrate these interconnections.

I then proceed to a discussion of the Indian situation. Chapter 2 reviews the principal features of India's irrigation development since Independence, the problems that have emerged, the responses they have evoked, and why they have not been successful. Chapter 3 focuses on institutional failures and goes on to spell out the directions of reform in three important areas: (a) planning and operation of large surface systems; (b) rehabilitation and expansion of local irrigation works to benefit rain-fed agriculture through integrated watershed development; and, (c) regulation of groundwater.

The solutions of course demand more and better technology; integrated planning and management of water with the river basin as a unit; combining large and small storages, and both of these with groundwater to minimize the adverse impact of large dams and provide users with greater flexibility; broadening the scope of water resource planning to include not just irrigation but also *in situ* soil and moisture conservation as a means of increasing the effective moisture availability for rain-fed crops. It is however important to recognize that technology alone is not enough; the effectiveness with which technology is used and its impact depend crucially on the way irrigation systems are organized and managed.

Irrigation organizations as they are currently structured are not informed by such a perspective. Water resource management is far too fragmented by type of works and in space; it is dominated by a civil engineering bureaucracy that is not only too rigid and narrow in its outlook but actively resists attempts to upgrade its technical competence and to broaden its expertise by inducting specialists from other relevant disciplines. A major internal restructuring of the government irrigation establishment is thus essential.

Privatization of such a basic, common pool resource like water and market allocation of this being neither feasible nor desirable, government must play a major role, but one that is very different from its current character. It needs to involve user representatives in system management and reduce its role in field level management by delegating a substantial part of the responsibility to user groups and creating incentives to induce them to assume this responsibility. At the same time, it needs to greatly strengthen its role in delineating a proper legal framework defining the basis on which rights of various claimants are to be determined, facilitating the participation of user representatives in management and setting up more effective mechanisms for regulation and conflict resolution.

While the broad directions of the necessary institutional reform are reasonably clear, working out its details and implementing them is far from easy. If the design of appropriate institutions in the face of the variations in environment, agrarian structure, and other related aspects is complex, engineering their reform is even more difficult. Invariably, and India is no exception, reformers do not have a clean slate to work on. They have to change pre-existing structures in terms of practically every feature relevant to harnessing and using water, and of interests, modes of functioning, and mindsets associated with them.

China is of special interest because its irrigation network is extensive, even as its characteristics and evolution differ markedly from those of India. Chinese irrigation developed earlier and was more extensive at the time of India's independence. It was mostly based on surface water. The systems are relatively small, and have grown by expansion and integration of older works constructed and managed by local efforts. This tradition continued after the revolution, with the local leadership comprising landed gentry being replaced by the party cadres leading the communes. Massive local mobilization for water conservancy work was a hallmark of the Great Leap Forward period. Problems of efficient management and cost recovery, and their solution, have been widely debated since the seventies. The abolition of communes in favour of individual cultivation in the post-Mao period has created new problems in water management. The Chinese experience shows that decentralized user management does not by itself solve problems of technology improvement, integrated management, and getting user communities and managers to work in harmony. The last essay in this volume provides a review of the history of irrigation in China, highlights developments in the post-revolution period, the problems encountered, and the attempts at reform.

This collection draws on my research extending over nearly two decades. Essays 1 and 4, completed and published in the early eighties, are reproduced here with minor changes in their original form. A postscript on the post 1979 reforms has been added to the essay on China. The two essays on India seek to gather the threads of work reported in various published and unpublished papers and weave them into an integrated whole. Taken together they reflect a certain evolution in my understanding of the problem. Thus, while reiterating the emphasis on viewing institutions as product of a complex, and evolving, interaction between the physical environment, technology, and socio-economic factors, I have come to feel the limitations of both the collective action theories and the techno-managerial approach in understanding them. The problem, therefore of striking an appropriate balance between the state and stakeholders in water management, the inherent complexity of evolving organizations to manage the task, the crucial importance of strong governance (in the form of a clear legal framework and strict enforcement) receive much greater emphasis in the essays on India. The ideas are not meant, nor claim to be, adequate or worked out in sufficient detail, but I hope they provide a framework that will help us to think about and devise meaningful reform for equitable and sustainable use of water in a broader, long-term perspective.

These essays are the product of a long process of learning and unlearning, changing perspectives and insights. The process has been helped and shaped in important ways not only by a reading of published literature, but by interaction with numerous scholars and people actually involved in planning and management of water, and most importantly, by my field research, while with the Madras Institute of Development Studies. It is impossible to mention all those who have contributed to my knowledge *and* understanding of water-related issues, but I would like to acknowledge a special debt to S. Janakarajan, S. Ramanathan, A. Rajagopal, K. Sivasubramaniam, and P. Anbazhagan who have been my collaborators in water-related research. My thanks are also due to P.S. Syamala for typing this manuscript with patience and care through my numerous, tortuous revisions, and to P. Anbazhagan and S. Hemalatha for helping check the final script.

Finally, redrafting part of the collection and editing the other papers was undertaken during 1996 and 1997 when I was a National Fellow of the Indian Council of Social Science Research. I would like to take this opportunity to record my thanks to the Council for granting me the fellowship and enabling me to concentrate on the preparation of this collection. I also owe much to the Madras Institute of Development Studies where I have worked since 1984, and am currently an emeritus faculty member.

Chennai
November 1998   A. Vaidyanathan

CONTENTS

CONTENTS

1

WATER CONTROL INSTITUTIONS AND AGRICULTURE: A COMPARATIVE PERSPECTIVE*

INTRODUCTION

Background

Irrigation institutions, and more specifically the relation between irrigation and general political authority, has long attracted the attention of social scientists and historians. More than a century ago Marx suggested that the apparent peculiarities of oriental society, which had been noted even earlier by classical economists like Mill, may have something to do with the technical and organizational compulsions of water control.[1] Weber also postulated a similar connection between the necessity for irrigation and the important role of the bureaucracy in ancient Egypt, West Asia, India, and China.[2] This line of argument was further elaborated by Wittfogel (1957)[3] into a general theory concerning the inherent tendency of hydraulic societies to become centralized, despotic states.

Social anthropologists interested in exploring cross-cultural regularities and the factors underlying them were naturally attracted by this hypothesis for, at a time when 'historians of culture were emphasizing differences between civilizations, Wittfogel was postulating a single basic factor that brought all these civilizations into being' (Steward, 1980). For the same reason it also provoked sharp critical reaction. The ensuing controversy stimulated a number of detailed studies of irrigation institutions in different parts of the world including Indonesia (Geertz, 1959; Jay, 1979), Ceylon (Leach, 1961; Chambers,

*Permission from IER. Wasteland News. (IJAE) for reproducing the material originally published there is gratefully acknowledged.

1977), Tanganiyaka (Gray, 1963), Iraq (Fernea, 1970), Central America (Price, 1971), Spain (Glick, 1970), Mexico (Hunt and Hunt, 1974), Spain and US (Maas and Anderson, 1978), Japan (Kelly, 1980), Thailand (Potter, 1971) and Taiwan (Pasternak, 1972).

These studies revealed that the institutional arrangements by which irrigation works are constructed and managed are extremely varied; that there is no systematic correlation between the existence of irrigation and the nature of the overall political authority; and that there is no general tendency for irrigation societies to be centralized, bureaucratic, or authoritarian. As Steward puts it:

The thirty years since Wittfogel's first publications... have produced a vast amount of field work which have thrown doubt on the universal applicability of the irrigation hypothesis. It is clear that in many instances irrigation has been ascribed excessive importance and that in others its development seems to have been the result rather than the cause of the growth of States. [Steward, 1980: 4.]

While there is impressive evidence to cast doubt on the Wittfogel hypothesis, we do not have an alternative one regarding the determinants of the form of irrigation institutions or their relationship to political authority. Part of the difficulty arises from the ambiguities in the concept of 'centralization' and of 'irrigation organization' that runs through most these studies. For instance, as Hunt and Hunt have noted, there is a tendency to confuse between two distinct types of centralization:

... One refers exclusively to authority in terms of the irrigation system. The other refers to generalized political authority which may involve other functions of control outside or above simple water control. In one case authority is exercised over different decision making rights in terms, exclusively, of the social and technical needs of the irrigation system per se. In the other case authority is exercised over water as one aspect of a complex political role or of a large multi-function political machine. [Hunt and Hunt, 1976: 132.]

This point is also emphasized by Kelly (1980: 14–20) who goes on to focus on the need for greater clarity in the concept of 'irrigation organization' itself. Irrigation, he points out, has several phases (namely control of the water source, the delivery of water, the actual application of water to crops, and drainage), each of which involves a number of distinct functions (namely, 'facility construction', operation and maintenance, water allocation and conflict resolution). In view of this, it is inappropriate, and certainly misleading, to speak of 'irrigation organization' as if it were a single entity handling all phases and functions; rather it has to be viewed in terms of arrangements for performing the various functions in each of the phases of irrigation and water control.

Moreover, as already mentioned, most of the above studies view water control institutions primarily in terms of their relation to general political authority. Though they document the wide variations in the internal structures and processes of these institutions (as distinct from their external form), there is scarcely any attempt to examine the significance of these variations or the reasons for these. It is of course recognized that institutions for water control have to be viewed in relation to agro-climatic conditions, the technology of water control and of agriculture, land tenure, and other factors that define the context in which these institutions functions and which to some extent condition their characteristics.[4] While most of the studies give some information on the physical and technological conditions of the study area, they do not always delineate the relations between them and the structure and working of water control institutions. This will entail systematic comparative studies of water control systems in different agro-climatic, technological, and socio-economic contexts. The need for it is appreciated (Downing and Gibson, 1974; Coward Jr. (ed.), 1980) and we do have some, though far from adequate, studies of the evolution of water control institutions in specific regions of China (Hamashima, 1980), Japan (Tamaki, 1979; Hatate, 1979, 1981; Kelly, 1980), and Taiwan (Vander Meer 1968; Pasternak, 1972) which throw light on these interrelations.[5]

In recent decades the role of water control in agricultural development has attracted greater and greater attention. That expansion and improvement of water control facilities (and in particular irrigation) has a crucial role in increasing agricultural production in densely populated developing countries is by now commonplace wisdom. Most countries in Asia attach great importance to rapid development of irrigation and flood control, and have spent massive resources for this purpose. Experience shows however that there is great deal more to this than the construction of reservoirs and canals. The effective utilization of these facilities is often found to be impeded by gaps in the terminal distribution and land improvement works, and lacunae in the organizations for maintenance and operation of these facilities on a continuing basis. This has stimulated interest in the problems relating to the design of water control institutions; what the 'right' design should be; and what the impediments to implementing the right design are. We now have substantial literature on these questions (see, for example, Coward Jr. (ed.), 1980; IRRI, 1978; Wickham, 1971).

Unfortunately much of this literature also tends to focus on institutions per se with what seems to be an excessive pre-occupation with differences in form, e.g. community managed vs. bureaucratic system; and centralized vs. decentralized systems. Not only is there no necessary correlation between form and effectiveness, but the appropriateness of institutional forms cannot be decided independently of the agro-climatic, technological, and land tenure conditions. This again stresses the need for, and value of, comparative and historical studies of irrigation institutions in a variety of situations.

The complexity of the task is obvious. It is nevertheless worthwhile, as a beginning, to attempt at least to sort out the key elements of the physical, the technological, and the socio-economic environment that have a bearing on the nature of the water control problem and hence on the institutions for handling it. It is to this end that this chapter is addressed. I hope it will help in the formulation of a more comprehensive and better-articulated framework in which comparative studies of water control institutions, their role and evolution can be undertaken.

The scheme of the chapter is as follows: since the ultimate purpose of water control is to help increase agricultural production, it is appropriate to begin by spelling out the relation between water control and agricultural production. This is the subject of the following section. We then proceed to deal with the role that institutions, along with other factors, play in shaping the construction of water control and distribution systems. Maintenance and operation of water control systems, both of which are aspects of continuing management usually handled by the same organization, are then reviewed. I have drawn a liberally on the available descriptions of water control systems in different parts of Asia (and more particularly India, China, and Japan) to illustrate the argument. The principal points of the discussion, which are suggestive rather than conclusive, are highlighted in the concluding section.

WATER CONTROL AND CROP PRODUCTION

Water serves two essential functions in plant growth: It maintains the plant temperature within tolerable limits and facilitates the absorption of nutrients. Plants can only use the soil/moisture available within their root zone. There is usually a certain maximum amount of water that a field of a given soil type and depth can hold at a particular point of time. When the soil moisture stock falls below this level, plants find it increasingly difficult to maintain the rate of transpiration necessary to regulate temperature. They do have internal adjustment mechanisms to cope with moisture stress without any adverse consequences to their growth. However, beyond a point, that varies with crop species and variety, continued stress begins to have adverse effects on the vital life processes of the plant, including its capacity for photosynthesis. These adverse effects increase as the moisture stress increases until in the extreme case the plant wilts and dies.

The amount of moisture available in the root zone also affects the volume of nutrients that the plant can absorb and the efficiency with which they are utilized: in general the more abundant the soil moisture, the larger the volume of nutrients that plants can absorb and the greater the efficiency of their use. Additionally, in the case of paddy, water plays an important role in regulating soil temperature and in controlling the growth of weeds.[6]

From the purely agronomic viewpoint, the 'ideal' soil moisture regime can be defined as that level at which the plant is free from any moisture stress and can realize its full genetic potential (which implies that other inputs are used in the required measure). Fields under crops are continually losing soil moisture on account of transpiration by the plants and evaporation from the exposed soil between them. Losses due to such evapo-transpiration (ET) must therefore be replenished promptly so that that the plants are not subject to moisture stress. It so happens that ET (under conditions of no moisture stress) depends largely on solar radiation, humidity, and other climatic factors.[7] Except in the case of paddy, the nature of the crop grown appears to make no significant difference.[8]

Typically, evapo-transpiration is relatively low in the winter, gradually rises to reach a peak in the summer months, and falls thereafter. In the absence of irrigation the only source of soil moisture is local rainfall. If the quantum and seasonal distribution of 'effective' rainfall (i.e. that part of the rainfall that is absorbed and retained in the soil) is such that the losses due to ET are promptly and fully replenished, year-round cropping is possible without any need for irrigation. The only limiting factor would then be the winter temperatures.

The above situation can obtain if the seasonal distribution of effective rainfall more or less coincides with that of ET, but this is rarely the case. As a rule, in most parts of Asia (barring the arid zones and deserts) moisture available from rainfall is on the average more than adequate to meet the water requirements of crops in the rainy season but falls short of it during the rest of the year. The relative seasonal profiles of ET and rainfall, which is one important manifestation of agro-climatic conditions, have a major bearing on the timing of sowings, the duration of the crop season, as well as the nature of crops grown under rain-fed conditions. The greater the seasonal concentration of rainfall and the larger the moisture deficit (i.e. the gap between ET and rainfall) during the drier months, the shorter is the period available for raising crops wholly on the basis of rainfall. The function of irrigation (and water control generally) is to bring the seasonal pattern of water availability more closely in line with that of ET.

There are several aspects to this: In the first place, even where, on the average, and taking a season as a whole, the amount of rainfall is adequate to meet crop water needs, its distribution within the season is often variable. Therefore, rainfall by itself can not be trusted to ensure that soil moisture is maintained at appropriate levels throughout the growing season. The variability in the date of onset of rains and in the amount of rainfall during the early phases of the season affects both the date of sowing and the rate of germination and establishment of the young plants. All this has a significant bearing on the eventual yields. Similarly, dry spells during critical stages of plant growth (which varies with crops) can reduce eventual yields considerably. Under these conditions, irrigation has an important role in ensuring that the soil moisture supply is adequate, especially during critical stages of crop growth, irrespective of the unpredictable variations in rainfall during the growing season. The lower the total rainfall, and the greater its variability during the season, the more important the role of irrigation.

In many regions of Asia, especially in South Asia, the nature of crops which can be grown during the wet season is also conditioned by the unreliability of rain and by the fact that the total quantum of precipitation is low. Thus, in most parts of peninsular India rain-fed lands can grow only millets, pulses, or oilseeds during the monsoon season. Under these conditions, irrigation works that augment the total supply of water during the rainy season enable farmers to grow crops like paddy which yield much more than other crops but require a larger volume of water. The impact is even more dramatic in arid regions like Rajasthan, Punjab, and much of Pakistan where rainfall is scanty even during the monsoon season.[9] In such cases the possibilities of rain-fed agriculture are severely limited: a relatively high proportion of land is under pastures or kept fallow, and the rest of it used to grow species and varieties of crops (mostly millets and pulses) which can survive on a small and precarious moisture supply but whose yield potentials are low. Introduction of irrigation in such a context enables a significant expansion of cultivated area, an extension of the cropping season, and a switch to altogether new and more productive cropping patterns.

Controlling soil moisture conditions, however, does not always or exclusively depend on irrigation: in many circumstances, the problem is one of protecting cultivated land from inundation and providing effective drainage in areas prone to waterlogging. Flooding arises from spells of heavy rainfall concentrated in a short duration and is liable to be especially serious in the plains along the lower reaches of larger rivers. Swamps and marshes in the estuaries of major rivers often require elaborate drainage works to remove excess water and make the land suitable for cultivation. (Extensive areas of cultivated land both in China and Japan were in fact reclaimed by this means.) Efficient drainage is also essential in all irrigated tracts to prevent waterlogging and salinity arising from indiscriminate irrigation. In view of these, the concept of water control relevant from the viewpoint of agricultural production should be broader than irrigation and also cover flood control and drainage.

It is clear that water control could contribute to increased agricultural production in one or more of several ways: It can raise yields of particular crops and make them more stable by facilitating planting at the optimal time and by enlarging the scope for fertilizer use; it can contribute to increasing the intensity of cropping by reducing the extent of fallowing and/or by extending the effective cropping season; and it enables a greater diversity of crops to be grown, permitting in the process a switch to high productivity, high value crops. However, the precise magnitude of the increase in productivity per unit area (and in total production) depends not only on the extent of the water control system and its quality, but also on several other factors, notably climate, soils, and the genetic characteristics of the crop varieties grown.

The more ample the rainfall and the more even its seasonal distribution relative to ET, the smaller is likely to be the increase in yields as a result of water control. The genetic characteristics of a crop variety would influence the extent to which the fertilizer response curve would shift when 'no moisture stress' water control is introduced. But in general, yield per hectare is likely to be higher at any given level of nutrient use, and the maximum amount of nutrients that the plants can use will be also higher under irrigated conditions than under rain-fed cultivation. Also, insofar as it makes for more abundant and assured supply of soil moisture at different stages of crop growth, irrigation also raises the responsiveness of yield to fertilizers. Increased productivity of fertilizers raises the economic level of application and therefore acts as a stimulant to faster growth of crop yields. Clearly, the magnitude of this effect will be greater under conditions of low and uncertain rainfall and when high yielding varieties of seed are used.

Given the agro-climatic conditions and seed variety, the ability of the irrigation system to regulate the water supply to individual pieces of land in its command and maintain an appropriate level of soil moisture at all stages of the crop growth makes a significant difference to the outcome. This ability, which reflects the quality of the system, depends upon the way the system is designed and operated.

The 'design' of an irrigation system has several aspects. These include: (1) the characteristics of the water source (which determines the quantum and seasonal pattern of water availability at the source as well as the limits within which the latter can be manipulated); (2) the distribution network and control structures (which determine whether or not water can reach all plots and how effectively, and within what range the timing and quantum of supplies to different plots can be regulated); (3) the quality of land preparation and the techniques of water conveyance and application (which affect 'irrigation efficiency', i.e. the proportion of water supplied which is effectively available in the root zone of the crops).

Within the limits set by the physical characteristics of the water source and the engineering design of the system, the way the distribution is organized and managed can make a considerable difference to the effectiveness with which crop patterns and water deliveries are regulated. Therefore, institutional arrangements for operating irrigation systems, i.e. the rules, procedures, enforcement mechanisms and personnel, have an important bearing on the eventual outcome. It is essential to recognize, however, that the organization of water management and the way it functions are conditioned by, as much as they condition, the natural and socio-economic context in which it operates. In the subsequent discussions I shall emphasize these interrelations a great deal.

CONSTRUCTION OF WATER CONTROL WORKS

The construction of water control works covers several distinct phases from controlling and harnessing a water source, through the laying of the distribution/drainage network reaching down to the farm plots, to preparation of land for efficient irrigation. The institutional aspects concern the location of the responsibility for planning, design, and construction of the facilities in each phase, for mobilization of the necessary resources, and resolution of the conflicts that arise in the process. Of particular interest in this connection is the relative roles of the state, local institutions, and private effort in each of these activities and the factors that determine the particular mix of roles in different situations.

Variations in Organization

There are marked differences between countries and regions in the manner in which the construction of water control systems is organized. In contemporary times the state everywhere plays a prominent role in planning, regulating, and assisting the development of irrigation, flood control, and drainage projects. The extent of its direct involvement in the process, however, varies. In India,[10] for instance, the national and the State governments bear a much greater direct responsibility than most other countries in Asia (including China): in the case of surface irrigation works it is the government agencies that undertake the preparatory surveys, design projects, and undertake the actual construction.

Till some two decades ago the government undertook the responsibility for constructing the main reservoir, the main and branch canals and distributaries up to outlets covering about 40 ha. The farmers were expected to construct field channels beyond this level and also institute the land improvements necessary for irrigation. There has however been a clear trend towards extending the government's sphere of responsibility to cover the construction of field channels, land levelling and improvement, and other on-farm works. The cost of works is supposed to be recovered from the beneficiaries in easy instalments over a period of several years in the form of 'betterment levies', but little is actually collected so that in effect the beneficiaries scarcely contribute anything to the cost of developing water control facilities.

In the case of groundwater, the state's role is more limited: apart from organizing surveys of groundwater potential, providing technical advice and some drilling equipment, and laying electricity transmission lines, its involvement in actual construction is confined to the relatively large-sized tubewells in north India. The bulk of the work of digging and drilling, installation of pumps, construction of channels, and land levelling is done by the farmers themselves, typically on an individual basis. However, the government is the principal source of finance, which is given directly (both in the form of loans and as grants), and indirectly in the form of loans through public financial institutions. This assistance is usually adequate to cover most of the initial costs and the terms are relatively soft.

In comparison to India, the higher levels government in China and in Japan play a much more limited role in practically every phase. The users and their organization bear a correspondingly larger responsibility both for construction and for mobilizing the necessary resources.[11] Thus in China

The planning, design, construction and operation of specific projects takes place at four different levels depending on the size of the project. The general principle is that the responsibility for a project which affects two or more units is taken up by the unit of the new higher rank. A corollary is that each unit which benefits from a project contributes labour and investment in proportion to its share of the benefit. Thus the central government, through the project bureaus of the Ministry of Water Conservancy, takes the responsibility for major dams and power stations projects, and labour to supplement state investment funds is contributed by all provinces that will benefit. Provincial governments are responsible for irrigation projects which affect more than one county or municipality. County (or municipal) governments usually undertake the diversions or reservoirs which affect more than one commune.... At the local level commune or production bridge units plan, construct and operate numerous projects of all types. [Greer, 1979: 116–17.]

Indeed, a high proportion of the water conservancy development works in China consists of projects of the last category. Until 1979 it was the explicit policy of the government that people's communes and their constituent units should bear the major responsibility for projects to harness water sources within the boundaries of the commune/brigade as well as construct distribution/drainage networks and undertake the land improvement necessary to make effective use of opportunities opened by larger projects undertaken by higher levels of government.

Even more striking are the differences in the method of resources mobilization. In China, unlike India, the beneficiaries are expected to contribute a large share of the project costs at the time of construction itself. Thus the cost of projects undertaken by communes/brigades/production teams are required to be met wholly out their own resources. Since labour is the principal resource needed for construction of these works and accounts for the bulk of the costs, the mobilization of resources principally takes the form of labour contributions by members of the beneficiary units. The communes are also expected to mobilize labour contributions for projects undertaken by the provincial/national government which benefit their areas. During the Great Leap Forward period such contributions were sought and obtained even for projects that did not directly or immediately benefit them.

Despite controversies over the basis of labour contributions and concern over the problems of effective organization of a highly variable labour force, the principle of beneficiaries contributing directly to construction has been maintained. Also, though labour contributions are now limited to units which directly benefit from a project, the magnitude of these contributions is very large, amounting in some cases to as much as 80 per cent of the total project costs (Vermeer, 1977: 260–1). It is relevant to note that the practice of mobilizing labour for both local and national projects for water control (and public works generally) has a long tradition in Chinese history.

In post-war Japan[12] the planning and construction of water control works, most of which are for the expansion and modernization of pre-existing systems, is the responsibility of the Land Improvement Districts (LID). The LIDs, of which they are 13,000 at present, are associations of farmers in the service area of water control systems (usually formed from users organizations of local systems built in the past) and managed by elected representatives of their members. They have been constituted under special legislation specifically meant to improve water control by constructing new storages to augment the water supply, rationalizing the layout of irrigation and drainage canals, and reorganizing the physical layout of plots to permit more effective water control at the field level and also facilitate mechanization.

The prefectural and the national governments undertake design and construction only of barrages (or storages) and canals which serve more than one LID. In all other cases the government provides technical advice (in matters of engineering and design) and financial assistance (half the cost is borne by the national government, one fourth by the prefecture, and liberal loan facilities are extended to cover the balance). While all this gives the government considerable influence over the general conception and design of the projects, the formal responsibility for decisions on all these matters, as well as the award of contracts rests with the LID management.

Historical Antecedents

To some extent these variations are a reflection of differences in the historical evolution of the government's role in relation to water resources development. In India, from what little we know about the irrigation history of the pre-British period, it would appear that considerable development of local irrigation works had taken place, especially in south India, under essentially local leadership even before the emergence of unified States of any significant size (Ludden, 1978). Similar developments may have taken place in north India too, but the state seems to have played a more direct and prominent role in developing some of the larger irrigation systems like, for example, the Jamuna and the Ganga canals and the elaborate system of flood irrigation in Bengal (Harris, 1923; Habib, 1963). During the early phases of British rule, the government took but a mild interest in irrigation development. In south India, for instance, soon after the British takeover, the government assumed responsibility for rehabilitating tanks and old systems like the Cauvery delta which had been damaged by war and neglect as well as their subsequent upkeep.[13]

In the first half of the nineteenth century the state spent substantial sums on reconstructing old systems of surface irrigation, some of them large even by modern standards. New projects were then taken up as commercial ventures. But since they were not profitable, progress was slow till the end of the nineteenth century. Thereafter, public irrigation construction increased rapidly, possibly as a response to recurrent famines and partly for strategic considerations.[14] Almost all the new projects were for surface irrigation; not only were they large in terms of cost and area covered, but many of them involved construction of reservoirs of a size altogether new to the subcontinent. With some exceptions, there was no attempt to extract contributions, whether in money or labour, from the potential beneficiaries or to involve them in any way in the conception or implementation of the projects. The post-Independence period witnessed a massive expansion in the scale of development in all types of irrigation and flood control projects. Surface irrigation works, almost all of them under state auspices, continue to be dominated by large storage works. Even in the case of groundwater, as noted earlier, the state provides technical assistance and the bulk of the finance.

Both China and Japan, by contrast, have had a long and sustained tradition of water conservancy development through local effort, and of requiring beneficiaries to contribute labour and materials for construction even when the government undertakes projects. In China huge projects to control floods of the Yellow and the Hwai rivers, and to divert river flows to irrigate extensive areas, were taken up as long ago as 500 BC or even earlier, and the practice of using corvée labour established then has continued throughout subsequent centuries. The role of such undertakings in shaping Chinese history and their significance for the nature and organization of government in China has attracted much attention. (Chi, 1936; Wittfogel, 1957.) The fact remains, however, that these large projects account for only a fraction of the total effort that has gone into the development of water conservancy works generally and irrigation in particular.

Not only was there a great deal of construction (smaller dykes, distribution networks, and land development) at the local level essential before the benefits of the large projects could be realized, but a substantial part of irrigation was, in any case, derived from essentially local works.

These works were conceived, planned, and constructed by, or under the leadership of, local landlords, clan leaders, and the gentry using local resources (mostly labour) with occasional assistance and support from the government and from merchants. Even in the case of large projects, mobilization of corvée labour, which on occasion could reach staggering dimensions, was widely used in project construction by both the provincial and national governments. Despite the far-reaching changes in other aspects of the economic, political, and administrative organization of Chinese society after the Revolution, one can see a certain continuity in the context of the organization and financing of water conservancy construction.[15]

The Japanese experience shows an unmistakable trend towards growing government involvement, not so much in construction, but in guiding and financing water control development (Tamaki, 1977; Hatate, 1978). Early development of irrigation consisted almost exclusively of small localized development organized by feudal chieftains and large landowners with the assistance of local labour. As the limits to such works were reached and/or significant improvements in techniques of water control become available, larger projects requiring the support of and intervention from supra-local political authority were taken up. This was sometimes the result of pressures from below, but often they also reflected the effort of ambitious local rulers to enlarge their revenues and extend their domain of power. Regional and national governments took a particularly active part in water conservancy during the seventeenth and eighteenth centuries, and financed part of the costs, but local participation continued to be very crucial.

Even during the nineteenth century, when the role of the national government increased significantly, the government generally insisted that the beneficiaries bear the bulk of the costs of new projects. It was only in the 1920s that the government, under the pressure of growing food shortages, substantially liberalized the scale and terms of assistance for water control projects.[16] For a variety of reasons, the process has been carried considerably further in the post-war period, but a high degree of user involvement in and control of the development activities has been maintained.

Scale and Complexity of Projects

To a considerable degree, differences in the role of the government reflects differences in the scale and complexity of the works involved. This is of course the central point of the Wittfogel thesis on the role of the state in hydraulic society. Large-scale, technically complex water control works require the mobilization of resources and organizational capacity on a scale far beyond the capability of local communities or private enterprises, and can only be undertaken by the government. This implies that where large projects and/or complex ones figure more prominently in the mix of water conservancy works, we may expect to find the state playing a more prominent role in their planning and construction. Indeed, such an association does appear to exist in the three countries just discussed.

Thus in India about two-thirds of the total irrigated area is estimated to be under surface water irrigation, the great bulk of it from canals as distinct from tanks[17] and other small, local sources. The canal systems, consisting of both diversion works and storage based projects, are typically large by the standards of the rest of Asia (except perhaps Pakistan). Projects irrigating 100,000 ha. or more are common and account for about half of the area served by surface irrigation sources. There are as many as 10 systems serving 500,000 ha. or more, and the largest, the Bhakra Nangal, irrigates some 1.3 m. ha. Reservoirs with a storage capacity of over 500 m. cu. m each, of which there are some 50, account for 80 per cent of the total storage capacity (147 bn. cu. m), while tanks and small ponds are estimated to account for less than 10 per cent (Rao, 1979). Large-scale canal irrigation projects dominated irrigation development during the period of British rule; in the post-Independence era, the absolute scale of activity has risen steeply[18] even though in relative terms their importance had declined as a consequence of even more rapid expansion of groundwater development.

In China, during the mid-fifties, surface irrigation accounted for over 80 per cent of the irrigated area, most of it (over 90 per cent) consisting of 'farm ponds and weirs' and 'gravity systems based on small ditches and aqueducts'. Large gravity canals, which would include the larger river diversion works (some of them of ancient origin) serve less than one-tenth of the country's irrigated area (Chao, 1970: 129). This underlines the dominance of relatively small, essentially local projects in the Chinese irrigation system and, as mentioned earlier, the construction of such projects, together with the improvement of existing local systems through communal effort, has been the highlight of the post-revolution period too.

It is in the case of flood control and drainage works that the central and provincial governments played and continue to play the major role. Many of the classic examples of massive public works in ancient Chinese history are in fact works of flood control (especially in the Yellow and the Hwai river basins), and navigation canals (like the Grand canal). In recent times too this has been the pattern. Besides the construction of new embankments and improvement of existing ones along the major rivers, there has been a large programme for the construction of reservoirs. The number of reservoirs with a capacity of 100 m. cu. m or more (most of which were meant for flood control rather than irrigation) rose from 10 in 1949 to 300 in the late seventies (Nickum, 1981: 6). The government had necessarily to play a leading role in the design and construction of these works which were not only huge but also required coordinated planning of entire river basins cutting across provincial boundaries. That even in these projects a large part of the cost was met by labour contributions from the areas benefited by them remains a distinct feature, and one that cannot be explained in terms of the scale and complexity of the projects.

In Japan, as in China, the bulk (75 per cent) of irrigated area is served by river diversions and ponds. Even now reservoir based systems are estimated to serve barely one-sixth of the irrigated area (Fukuda, 1976: 88).[19] The Japanese irrigation systems are large in number and their average size is quite small. If we consider each Land Improvement District as one system, the average scale of an irrigation system is barely 250 ha.; there are hardly any systems serving over 20,000 ha.[20] Flood control and drainage were of course essential for expansion of paddy cultivation and in some areas (like the Kanto plain) involved relatively large-scale construction and a high level of technique. However, nowhere did these works reach the scale found in China. Much of it could be and was handled by local rulers well before the emergence of a unified national government. The relatively more prominent role of the government in the post-war period was consistent with the need for construction of new storages, involving integration of pre-existing irrigation communities and LIDs.

Given that the scale of projects has some bearing on the nature and extent of state involvement, the question arises as to what determines the type of projects that are undertaken. The nature of water control needed for agriculture depends on agro-climatic conditions, while what is feasible is conditioned by topography, geological conditions, and the state of the art in hydraulic engineering; and what actually is selected from the feasible set is a function of yet other, essentially socio-economic, factors.

Agro-Climatic Factors

I have already explained at some length how rainfall and evaporation are the two basic climatic variables that determine the nature of the water control needed for agriculture. Different configurations of these variables imply different problems in the attainment and maintenance of appropriate soil moisture conditions. Consider for instance the two different configurations shown in Fig. 1.

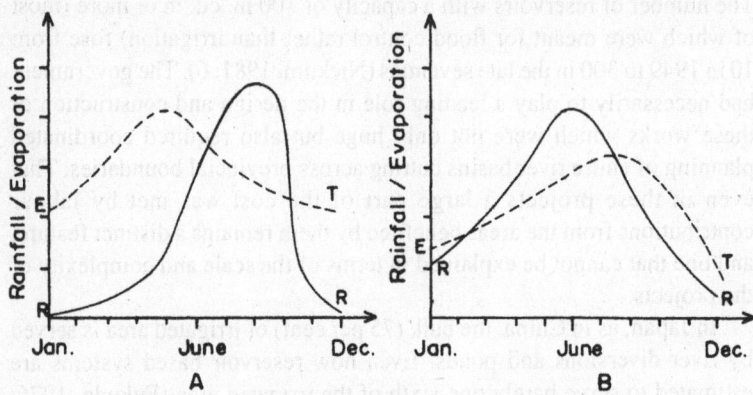

Fig. 1: Climatic Patterns Typical of South and East Asia

Note: A = Stylized Pattern in South Asia, B = Stylized Pattern in East Asia.

In A, which is a stylized depiction of the climatic pattern characteristic of much of South Asia, temperatures are throughout much higher than in B, which approximates the climate characteristic of much of East Asia. Consequently, A has higher evapo-transpiration (ET) in practically all seasons and certainly over the year as a whole in comparison to B. On the other hand, the average rainfall in A is lower and its seasonal concentration much more pronounced. The combined result of these differences is that the 'dry' season in A is not only longer but the moisture deficit (i.e. the excess of ET over effective rainfall) is also larger. Under these conditions, the irrigation needs of the dry season, being relatively large in relation to crop water needs, can be met only if the surplus water from the monsoon season is stored either on the surface or underground for use during the dry season. Which of these possibilities is in fact available, however, depends on topography and geology.

Thus in the Indo-Gangetic plain, though the general climatic pattern follows pattern A, the rivers flowing through the plains have unusually large catchments which includes the Himalayas. The contribution of snow melts from these mountains is an important factor in making all the major rivers in this region perennial. The topography of the plains also permits irrigation by diverting the river flow to feed canal systems.[21]

Apart from the Indus valley systems of antiquity, several large canal systems based on run-of-the river were constructed in this region both before and during British rule. However, as the possibilities of river diversion were exhausted, further expansion has increasingly depended on storage.[22] The geology of the plains is exceptionally favourable for groundwater storage, but the intensive exploitation of this resource had to await the introduction of energized pump-sets (which reduced the cost of lifting water) and the availability of techniques for tubewell construction (which made it possible to tap deeper strata of the sub-surface storage).[23]

In south India, by contrast, the rivers are mostly seasonal; there are no extensive plains along the course of the major rivers; and geology is not favourable for groundwater storage. Taking advantage of the fact that the region gets rain from both the monsoons, local topographic variations have been exploited to impound rainfall in an extensive system of tanks which are used to raise irrigated paddy and simultaneously serve as a means of improving groundwater recharge in their command area.[24] Barring diversion works in the deltaic regions of the major rivers (the Krishna, the Godavari, and the Cauvery, and to a limited extent in the smaller river basins), the bulk of the irrigation in south India in the pre-Independence period was from ponds and wells.[25] The development of tank irrigation appears to have reached a point of saturation even before the British came. Further significant expansion of surface irrigation in this tract required the construction of large storages upstream and/or less expensive techniques for lifting groundwater. The technology required for storage dams did not develop indigenously, and became available only in the last century or so. Economical techniques for exploiting groundwater did not become available until much later.

By contrast, in East Asia (pattern B) precipitation is higher and at any rate better distributed in relation to ET; the duration of the season when there is a soil moisture deficit (under conditions of rain-fed farming) is shorter; and the magnitude of the deficit in relation to crop water needs is smaller.[26] Rainfall, being more evenly distributed, most of the rivers are perennial. In China, the catchment area in relation to arable land is enormous and the rainy season in the upper parts of the catchment is somewhat earlier than in the lower part of the river basins (where most of the agriculture is concentrated). These factors probably contribute to smaller seasonal variations in river flow in the lower reaches. In this conjunction of circumstances, it is possible for China to meet the dry season moisture deficits with the aid of diversion works, ponds, and surface water lifts. Both in China and in Japan large canal irrigation systems, and in particular those based on storage, are a relatively recent phenomenon.

By the same token, the relative abundance of water in these countries makes floods a more serious problem. Effective measures to regulate seasonal floods and ensure proper drainage of excess water is essential, especially in the lower reaches of river basins and the estuarine areas. These works tend to be relatively large both in terms of size and the extent of area benefited in comparison to irrigation projects. The flood protection works of north China are among the largest water control works in the world.

Technology of Water Control

The constraints imposed by the state of hydraulic engineering and construction technology have been greatly relaxed in modern times because a large accumulation of proven techniques is available and has become more easily accessible. These constraints were however quite important in earlier times. Thus the level of engineering and construction skills required for large reservoirs and extensive canal systems is of an altogether different order in comparison to what is involved in the construction of local ponds and shallow wells. For reasons that are far from clear, there was significant progress in technology in some situations and at certain times but not in others. Thus, in China sustained efforts to control her major rivers led to major advances in techniques of flood control fairly early, and this made it possible to implement massive works even in the pre-christian era (Needham, 1971). Corresponding improvements in storage dam construction and groundwater exploitation did not however come about even in situations where the potential for such works existed and the limits to development on the basis of simpler techniques had been reached. That the technology of storage based systems did not develop indigenously in India and did not become available till recently must have been one of the factors that arrested the growth of irrigation in this country.

The relevant technology of water control from the perspective of agriculture of course extends far beyond what is used for harnessing water sources. It encompasses techniques for regulating conveyance of water and its efficient application to crops. Here, perhaps more than anywhere else, one can see the close interconnection between the physical and the organizational aspects of water control. Efficiency of water management is crucially dependent on the working of institutions operating the system, and within limits may even make up for defects in system design. However, (1) *the physical design of the system (embodied in the structures through* which the flow of water to different parts of the system are regulated; (2) the specifications, design, and layout of the irrigation–drainage channel network; (3) the irrigation–drainage channel network; (4) the quality of land preparation; and (5) the irrigation technique used at the field level) set definite limits within which the institutions have to operate. Besides, it is also relevant to note that the notion of 'efficient' water use is itself conditioned by the state of knowledge regarding water–yield responses, which again has been constantly changing. There is a striking difference in this regard also between east and South Asia. Most of the Japanese and Taiwanese systems, and some at least of the Chinese, have been 'modernized' and achieved a level of sophistication in the design of distribution networks and control devices, as well as the management of water deliveries, that stand in marked contrast to the south Asian systems.

Economic and Social Factors

Since water control is but one of the ways of increasing agricultural production, the relative costs of alternative means of achieving this objective is clearly relevant in determining the extent of resources devoted to water control and what kinds of water control facilities, within the limits set by technically feasibility, are to be developed. When there is an abundance of uncultivated, but cultivable land, the need for water control would be less pressing. Also, if extension of cultivation is cheaper in terms of the resources required per unit of additional production, we would expect greater emphasis on bringing new land under cultivation than on intensifying cultivation by extending and/or improving irrigation. That expansion and improvement of water control as a basis for intensive irrigated agriculture occurred earlier and has been carried much further in China and Japan than in India perhaps reflects the fact that the former reached the 'frontier' of extensive cultivation much earlier, thus making more intensive cultivation based on proper water control essential, and probably more economical, for raising productivity.

A similar explanation has been suggested for differences in the timing and scale of different kinds of agricultural investments between Japan, Korea, Taiwan, and the Philippines (Kikuchi and Hayami, 1979.) Clearly, the relative costs of and returns to different ways of raising production are a function of the state of technology for land reclamation, hydraulic engineering, lifting of water, and in plant breeding and agronomy. Consequently the emergence of new techniques in any of these fields affects the choice. It is not an accident that ground water irrigation spread rapidly only after energizaed pumping and tubewells, both of which greatly reduce the cost of groundwater, came into the picture. Similarly, the introduction of new high yielding, fertilizer responsive crops and varieties, and of cheaper sources of plant nutrients have raised returns to irrigated agriculture. This has in turn stimulated larger investment in water control.

However, the economic calculation is not always explicit. Indeed, there is little evidence of conscious calculation behind much of water conservancy development in history. Systematic economic evaluation is a relatively recent phenomenon. While project evaluation is now widespread, the basis of the economic calculation is not only complex but strongly conditioned by extra-economic factors rooted in the way resources are mobilized, the interests of the group(s) that mobilize the resources, and the distribution of costs and benefits amongst the beneficiaries.

The calculation is relatively straightforward when the proposed investment is made and operated by a single farmer exclusively for his benefit. It is a reasonable presumption in this case that his decision will depend on whether or not he will get a better return from that investment than from other alternatives available to him. There are indeed examples of such individualized private investments in water control. Shallow wells and pump-sets typically fall in this category. However, the more common situation is one in which an irrigation project benefits a large number of users and has necessarily to be constructed and operated by, or for, the group of farmers.

Consider for instance investment in a new tank or modernization of an existing village tank or the construction of field channels, land preparation, and other works to make effective use of available water in a small community of farmers. The fact that each of them acting individually cannot go far in exploiting any of these possibilities and that combined action would be to everyone's benefit implies that there exists a certain mutuality of interests as a basis for undertaking the investment collectively. But this alone is insufficient to induce the group to actually undertake the project. Even if the overall return to investment is attractive, whether or not they do so depends crucially on the individual beneficiaries' assessment of their respective net gains.

There are many reasons why individual beneficiaries may be unwilling to participate. For instance, the modification of an existing system may be collectively profitable, but farmers and farming communities served by the existing system could have apprehensions that the change will adversely affect the quantum, timeliness, or reliability of water supplies which each of them now get. These apprehensions have to be allayed and overcome before the collective investment can be made. This involves demonstrating that the new investment will not only augment total supplies, but guarantee existing users' supplies and/or recognize the superior claims of existing users in water allocation under the modified system.

Another reason for reluctance to participate in collective effort could be the uncertainty of net gains to the individual. The gains to the individual farmer are determined by several factors, some of which have to do with the way the costs of and benefits from the investment are shared among the beneficiaries; some with the location, soil, and other characteristics of particular plots belonging to them; and others related to the command (in terms of magnitude and quality) over complementary resources necessary to take advantage of the improved water control

The first of these has to do with the principles and procedures concerning contributions to project construction/maintenance from, and location of water in relation to, individual beneficiaries. Typically, collective investments do have such rules. But the mere formulation of rules and procedures is not enough. Also relevant is the individual user's assessment of how effectively and impartially these rules will in fact be implemented. If some or all the beneficiaries feel that some users, by virtue of their (upstream) location and/or their power over the organization/ personnel managing the system, could, and are likely to, manipulate the allocations to the latters' advantage, the inducement to participate in collective effort will be weakened. The inducement could also be weakened because an individual falling in the service area of a common investment can secure his needs without getting involved in the collective effort, either because he will get the benefit anyway or because he can independently develop supplementary sources (like wells).

Even where the rules are clear and everyone is confident about their application, there remains considerable uncertainity about costs and benefits to individuals arising from inadequate knowledge about the impact of the investment on account of the wide variations in individual circumstances with regard to land quality, access to inputs and know-how, and the ability to afford the necessary inputs; and from uncertainties in the timing and quantum of water supply in any particular season. The land tenure status of the individual has also a bearing. An owner–cultivator's assessment of the benefits of joining a collective water control project will obviously differ from that of a tenant. Among tenants, the form of tenancy contract and expectations about the effect of improved water control on land rents is likely to influence the individual's cost-benefit calculations.

While little can be done about the uncertainty in the overall supply of water in the system or its seasonal pattern, the existence of well defined rules of allocation under different contingencies and the confidence that the actual allocations will conform to rules would make a significant difference. The latter in effect depends on the credibility of the functionaries of the irrigation organization and of the ability of its 'leaders' to discipline violators of rules and enforce compliance.

Traditionally, the involvement of dominant landowning classes who stood to benefit directly from the water control investment or other centres of local power (like the gentry in China) has been one way of giving strength and credibility to irrigation organizations. In many cases the bulk of the benefits went to these classes, who in effect used their power to get the more numerous, but lesser, beneficiaries to fall in line with their programmes and force than to contribute the development of the system. Examples of strong organizations comprising farmers of more or less similar economic background are to be found in Japan and Taiwan, but elsewhere they are rare. Even in the case of the people's communes in China—notwithstanding communal ownership of land, a relatively egalitarian distribution of benefits and collective decisions on development and use of available resources—consensus on these questions has been by no means easy or conflict-free. Often it has required strong and active intervention by the Party cadres, which have been until recently the principal loci of local power in Chinese rural society, to arrive at and implement decisions (Anon, 1958; Crook and Crook, 1962; Nickum (ed.), 1981).

The problem becomes more complex when the scope of water control works extends beyond individual communities, and when villages lose their isolation and become increasingly integrated with the economy and polity of larger territorial complexes. The former means that coordination and resource mobilization have necessarily to be managed by a supra village authority. The latter has the effect of loosening, and making more indirect, the connection between the mobilization of 'surplus' and its utilization for water control. The extent and manner of mobilization as well as its disposition is then naturally subject to the influences of the configurations of interests and power over a much wider area and further away from the villages.

Historically, investments in water control were promoted or undertaken by supra-village authorities, eager to extend their command over territory, resources (by way of rent and taxation), and hence, power. In some cases, like China, these authorities had the means of enforcing mobilization of corvée labour and the capacity to organize them for the construction of large-scale works. More often however they preferred to encourage (by way of concessional taxation for instance) merchants and large landowners (or their retainers) to invest in agriculture (including the construction of water control works) rather than commit their own resources directly. Examples of this can be found in virtually every country (see, for example, Kelly, 1980; Ludden, 1978).

I have already highlighted the fact that the extent and nature of involvement of government in water control development varied a great deal between countries and even within countries at different points in time. This does not admit of any simple explanation. One can however list some of the more important factors that would figure in any explanation. They include: the magnitude of resources at the government's command (largely a function of the land revenue system and the effectiveness of its administration); the land tenures (which influence how the costs and benefits of water control are to be shared among different layers of society); the necessity for and possibilities of investing in water control; and the claims of other uses on resources available to the government.

The picture has changed dramatically in recent times with states' involvement in water control becoming increasingly widespread and important. This reflects in part the growing scale and complexity of projects and the larger resources required, greater expertise, and a higher level political mediation of conflicts between sectoral and regional interests. In part this trend also reflects the heightened pressures for accelerating improvements in mass living standards and the acceptance by the state of its responsibility to mobilize and use resources for this purpose.

An inevitable consequence of this growing centralization of mobilization and use of surpluses is the intensification of competition among regions and classes for available public funds, and the consequent enhancement of the role of the political forces (i.e. the relative power of the competing claimants) in deciding allocations. There are however significant differences between Japan, China, and India in this regard. Thus in Japan, growing state involvement has taken the form of the national government meeting a large part of the costs (including liberal subsidies) and providing the legislative framework leaving much of the actual decisions and implementation to autonomous irrigation organizations. In China, the state's financial contribution to irrigation development has been consciously limited, but the power of the state is used via the party cadres to facilitate collective action at the project level.

By contrast, the compulsions of electoral politics has led the Indian state to take over a much greater degree of direct responsibility for planning, finance, and actual construction of irrigation works, including those components of it that can be quite effectively undertaken by beneficiaries at the local level. In the process the inherent conflict between 'efficient' use of available resources and the propensity to use public funds as a means of receiving and widening electoral support has been greatly aggravated.

Faced with a demand for expanding irrigation far in excess of what can be accommodated with available resources, governments have responded by starting too many projects, many of them without proper investigation, and spreading the available funds thinly among them; permitting canal systems to be extended far beyond the limits of the area that can be served by the available supplies; and a striking reluctance to face conflict-ridden problems of consolidation of holdings, land improvements, localization of the command area for different crops and collection of betterment levies. The resulting delays in the completion of the works and serious imbalances between water supply, irrigable area, and crop patterns in the command area of completed projects, increase the conflicts over water allocation and add to the uncertainity of water supply, its cost, and productivity.

OPERATION OF WATER CONTROL SYSTEMS

Unlike the construction of water control works, which is essentially an one time or, at most, an intermittent activity, their operation involves tasks of a continuing nature. Basically these tasks are to ensure that the physical facilities (dams, canals, field channels, structures) of the system are maintained in good working condition and to regulate access to and use of the facilities provided by it.[27]

When a water source is harnessed and used by a single farmer to culivate his own land, he can decide how best the available water is to be allocated between different uses (crops/seasons) in relation to his objectives. The question of allocation between users does not arise and, subject only to technical constraints, he is fully in control of implementing the decisions. The problem remains fairly simple even when the irrigation source supplies water to several users, whether because the supplies are in excess of the owner's needs or because the source is developed purely as a business proposition, provided the owner has unfettered right over the disposition of the water. The best examples of such a system are individually owned wells or tubewells. Some large cultivators had, and still do, have ponds and small diversion works for use on their own land.[28]

However, as typically in most parts of Asia, even small surface systems serve several farmers so that, the allocation of both maintenance obligations and of water between users is as important as that of allocation between uses. The mechanisms and procedures for deciding the allocation rules and enforcing them, which together define the organizational structure for operation of irrigation systems, are therefore of great importance.

Organizations for operation of water control systems tend to be differentiated in terms of the role of bureaucrats (as distinct from users) in management and the extent of centralization of authority. 'Bureaucracy' refers to the corpus of paid professional staff hired to carry out specified tasks in an organization within the framework of certain recognized rules of procedure. Small localized systems serving a few farmers can make do with a simple organization and manage all the tasks with the assistance of their own members. This is in fact the case: in most of Asia such systems traditionally do not use any hired personnel; both the policy-makers and the administrators are chosen from within the community by selection, election, or rotation. That the personnel are not professionally trained, however, does not mean that they are not skilled. Indeed, considerable knowledge and experience in handling both the technical and social problems of water management are essential for smooth functioning of even small systems, and the functionaries are expected to have these attributes.

Larger, multi-community systems require more skilled and specialized personnel to manage the technical tasks, and also personnel who can give greater continuing attention to the day-to-day tasks of running the system. Consequently, as the size and complexity of the system increases, the need for paid, full time staff to handle both technical and routine administrative tasks also increases. There are however situations in which the bureaucracy's role goes beyond technical and administrative functions to comprehend making and enforcing policy. Therefore, it is not so much reliance on paid, full-time personnel as the role they play together and the way in which they are appointed and controlled that forms a meaningful basis for distinguishing different classes of irrigation organization. In this regard too there is a marked contrast between the patterns prevalent in east and South Asia.

Thus in Japan, because of the way irrigation systems evolved and the predominance of small, localized systems, user control over the management (with regard to policy formulation, day to day administration, as well as appointment and control of operating personnel) is very strong. As systems grew in scale and complexity, reliance on paid professional staff also increased. With few exceptions, however, all the staff are effectively under the control of the Land Improvement Districts (LID) that are managed by representatives of users. Apart from laying down the general legal framework governing LIDs and, in rare instances, managing reservoirs and canals serving more than one LID, the state plays a relatively marginal role in the operation of irrigation systems.[29]

In China,[30] although from ancient times the state bureaucracy was supposed to be responsible for managing water control works, in actual fact its role was limited to supervising the maintenance and operation of large projects constructed by the state, and that too to the central reservoirs, canals, and embankments. Lower levels of large systems, as well as smaller systems constructed with local resources and initiative, were managed locally with their own personnel. Currently, for purposes of operational management, multi-commune water control systems are organized into Irrigation Districts of varying sizes ranging from a few villages to several communes or even prefectures. All these organizations have a corpus of paid professional and administrative personnel appointed and controlled by the Irrigation District. But they work under, and are answerable to, management committees which include representatives of users.

In India[31] too, user management is conspicuous in the case of ground water (which is mostly exploited through individually owned well and pumps) and in traditional local systems. Management of canals however (comprising the bulk of surface irrigation) is dominated by and dependent on government bureaucracy. Throughout India the operations of surface irrigation—from the source, through the main and branch canals, up to outlets commanding around 40 ha.—is wholly organized and manned by government personnel. The overall responsibility for management of each system is vested in a senior official almost invariably drawn from a permanent cadre of engineers.

The entire project command is divided into a hierarchy of smaller operational units and sub units (usually by sections of main canal, branch canals, and groups of distributaries) under officials subordinate to the system manager. The intermediate level managers are again drawn mostly from the state cadre of engineers. Their support staff, including those responsible for supervising and policing operations at the distributory/ outlet level, are all employees of the government. Matters of general policy are decided by the state irrigation departments which formulate detailed regulations by way of statutes, rules, and operational manuals to guide managers of individual systems.

This pattern has been adopted even in areas with a prior tradition of irrigation and local institutions to manage it. The latter, to the extent they exist, take care of essentially local matters, occasionally pressing local interests (for more water or better regulation of it) with the system management or the government but never as an integral part of it. The management of the Indian canal systems thus represents a high degree of centralization in the sense that the authority to make and implement policy and to resolve conflicts vests in a bureaucracy answerable to the state.

It would appear useful, in the interest of clarity, to draw a distinction between two aspects of centralization of irrigation management: one might be termed the 'functional' and the other the 'political' aspect. The former reflects the technical characteristics of the system that dictate the locus and the distribution of decision-making powers between different levels. For instance, a system that is entirely based on canals fed by a single source (say a diversion work or a reservoir) requires centralized coordination of the operation of canals and outlets, and once the operational schedules of the canals are determined, the amount and timing of water available at each outlet is more or less fixed. The management organs at the outlet level can protest against upstream users utilizing too much or can try to take more than they are entitled to, but in either case, the extent to which they can manipulate either the timing or quantum of supply is limited. Their principal function would be to distribute supplies available at the outlet between different parts of the area covered by that outlet.

By contrast, a canal irrigated tract that combines the main storage with smaller local storages/wells widely distributed in the command area, offers much greater scope for manipulation of the quantum, timing, and allocation of water at the intermediate level. This means that potentially the intermediate management levels have a wider range of decisions to make and implement. The extent to which this potential is in fact used however depends on several factors, including the adequacy of irrigation water supply in relation to the requirements of the command area.

The 'political' aspect concerns the question of who among the various groups interested in the system makes the allocation decisions, on what basis, and for whose benefit. The allocation decisions have to be derived from certain objectives (which may or may not be explicit) to be achieved by the system. Since every allocation decision in a multi-user system implies a certain distribution of costs and benefits among the users, its smooth functioning requires clarity regarding the goals (that must necessarily comprehend both the overall level of output and its distribution among different parts of the system), definition of allocation rules consistent with these goals, and the capacity to enforce them. It is conceivable that these matters can be settled by mutual agreement among the users or their representatives. But where there is no consensus among the users or between the users and the management on these matters, or the consensus is fragile, it will be difficult to enforce the allocation rules. Smooth functioning of the system requires that there be some higher authority to enforce rules and resolve conflicts.

This aspect of centralization of authority is noticeable even in traditional user-managed systems. The most obvious case is when the dominant landholders are also the managers. It is not however necessary for the dominant landowners to be directly involved in water management; they may exercise authority through managers who are their nominees or subserve their interests. They may also choose to interfere only when the conflicts threaten to get out of hand. Traditionally, the authority of large landowners appears to have played an important role, at any rate as an authority of last resort, in the functioning of local systems. When the composition of this group changed, as it did periodically, the working of the irrigation organizations was also affected. There are cases, of which the Vel Vidanes of Sri Lanka is an excellent example, where the overriding authority was formally vested with a functionary backed by the state, but they again seem to have been drawn from, or into, the ranks of the well-to-do.

In multi-community systems, large landowners per se, appear less likely to play the role of 'ultimate arbiter' because there are so many more of them and conflicts of interests among them are apt to be greater. And where, as in modern Japan, China, and Taiwan, land reforms have resulted in a relatively even distribution of land, landownership is much less important. There they rely more on rules and adjudication procedures, and on entrusting the implementation to persons recognized by the community to be knowledgeable, skilled, and fair. Even then, there is need for a clear locus of authority for resolution of disputes which the organization cannot cope with in the normal course. In several systems, the top level functionaries are people who command respect and influence in the region on the basis of wealth or political influence.[32] In the case of China, a strong and active involvement of Communist Party cadres performs the same function.[33]

On the other hand, one cannot assume that merely because the management of irrigation organization is centralized and vested with wide powers, that the central authority is in actually, or can be, effective. Thus, as we shall see later, in the Indian canal systems the irrigation bureaucracy is not free to exercise its very considerable formal authority because the rules do not reflect or derive from goals on which both users and managers are broadly agreed. Attempts on the part of the bureaucracy to enforce its rules are thwarted in a multitude of ways.

Since there are so many dimensions to irrigation organization, it is difficult to find a classification that is simple and at the same time meaningful. In any case, the form of an organization tells us relatively little about how it actually works and how it adapts itself to changing circumstances. These questions are best examined in relation to the way specific tasks of management in irrigation works, namely maintenance of physical facilities and regulation of water allocation, are dealt with in concrete situations.

MAINTENANCE OF IRRIGATION SYSTEMS

The purpose of 'maintenance' is to ensure that physical facilities (dams, control structures, distribution networks) function smoothly and at the level of performance for which they were designed. Typically, in a surface water system this involves periodic inspection of the facilities to identify any deterioration (such as leaks in embankments, erosion, silting of canal beds, growth of weeds, malfunction of sluices, etc.) and execute the necessary repairs. Besides, the organization needs to be alert in identifying major malfunctions as they arise and have the capacity to correct them promptly.

Inefficient maintenance could adversely affect water deliveries by reducing the volume of water carried by the canals; slowing down the speed of water flows; increased waste due to leakage and spills; and, in extreme cases, a partial or total breakdown. All these reduce the volume of water made available to the fields and hence the feasible level of production in relation to the potential of the system. The quality of maintenance affects the interests of both the organization as a whole (which is presumably interested in getting the maximum production with the available water) and the users (whose output and incomes are directly affected by it).

The strength of this common interest is however variable. It seems likely to be strongest and most widely shared in relatively small systems which has been set up by the community of users who have also made a substantial contribution to the cost of developing it. The same is true in systems where neglect of maintenance leads to heavy loss in productivity, and when there is an adequate supply of water to meet the needs of the entire service area. On the other hand, the larger and more extensive a system, the more difficult it is for users to appreciate their common interests in proper maintenance of facilities at all levels. The importance of maintaining the main canal is not so obvious to all the users of a system serving 100,000 ha. as in one serving 100 ha. The interest will naturally be weaker when the users have not contributed any of their own resources to developing the system and when the returns to irrigation in terms of increased output is relatively small or uncertain. If all parts of the system are not supplied the promised quantities, and if the supplies are irregular and uncertain, not only will the affected segments have less interest in contributing to system maintenance, but they may actively resist levies for this purpose as being 'unfair'.

Traditional community systems of Asia, being small and generally the product of local effort, fall in the first category. From detailed descriptions of several such systems, it is clear that they usually have established conventions regarding the timing of repairs, the way in which the work is to be divided, the responsibility of functionaries at different levels, and the obligations of users. While the force of custom and social pressure facilitates the smooth working of the arrangements, it would be wrong to suppose that they work entirely on the strength of mutual interest reinforced by custom. Most systems find it necessary to specify sanctions (ranging from fines to loss of water rights) against non-compliance and to have a centre of authority that can enforce the sanctions to ensure compliance.[34]

Historically, the centre of local power—generally, though not always, the large landowners—had an important role in this regard. It is to be expected that the interest of large landowners in the management of water control would be influenced by the extent to which their incomes were affected by the way the system was managed. The active interest taken by large landlords of Japan in water control management may have something to do with the fact that they lived in the village and cultivated a part of the land directly.[35] In south India the traditional land tenure system earmarked a certain percentage of produce for maintenance of tanks and the land controllers took active interest in this task. In other parts of this region, a large number of tanks were owned by landlords. So long as their incomes (as owners and as lessors) and the state's revenues were linked to the total production on irrigated land, there was strong interest in both local and supra-local centres of power in ensuring proper maintenance.

Wars and tenurial reforms early under the British rule weakened these traditional centres of village authority. This undoubtedly contributed to the deterioration of irrigation tanks during the nineteenth century.[36] Though the government took over responsibility for maintenance, its functionaries were unable to get the work done.[37] In Bihar too the traditional zamindars, who had taken an active interest in maintaining local irrigation works, lost interest once the share rent system was replaced by a fixed rent system (Sengupta, 1980). Clearly there is a close correlation between land tenure and the management of local water control systems.

In larger systems, the functions are of necessity more diffused and the perception of common interest in the maintenance of the system as a whole tends to be weaker. Consequently, the rules and procedures also tend to be more formalized and impersonal. The mobilization of labour contributions become more difficult. That is perhaps why some of the large multi-community systems combine a labour contribution by beneficiaries for maintenance of local facilities with a levy to meet the cost of maintaining facilities that fall under the jurisdiction of higher levels.[38] Where, as in China, maintenance of even large systems was sought to be effected with labour contribution, support of large landowners, the gentry, and other centres of local power was necessary to effect the mobilization. The quality of maintenance was apt to suffer when government as a whole was weak and/or there were serious conflicts of interest at the local level.[39]

The growing difficulties of enforcing corvée labour gradually led to the abolition of the system. By the nineteenth century the cost of maintaining public works was supposed to be met out of the general budget. However, the bureaucracy was not allocated sufficient funds to carry out the works for which they were supposed to be responsible. In some areas officials resorted, with the help of local leaders, to extra-legal labour contribution. There were attempts to narrow down the state's responsibility for maintenance. In many cases officials simply neglected this task. The general decline of the authority of the central government was undoubtedly an important factor contributing to the relative neglect of maintenance of water control facilities under state control during the decades preceding the Revolution.

With the reorganization of water control management in the post revolution period, labour contribution for maintenance has again become important. Almost all the labour needed for maintenance and repair throughout the system is contributed by the beneficiaries. The requirements of items other than labour, as well as the cost of the administration, are recovered in the form of water fees. While the responsibility for planning and executing maintenance works is distributed between different levels, and local leadership is encouraged, the available accounts (mostly relating to models of successful management) highlight the key role of the party leadership at all levels of the organization through a combination of propaganda, political education, and even compulsion, in ensuring smooth working of the arrangements. There are indications that when the party leadership is not strong or, as it happened after the 1979 reforms, the Party involvement is reduced, the functioning of the water control organization is adversely affected. (For details see Chapter 4.)

In India (and South Asia generally) the state has direct responsibility for maintenance of all major canal systems and a large part of the tanks and other local systems. Traditional arrangements, whereby maintenance is done locally by mobilizing labour and other resources (including income from common properties designated for the purpose), are still to be found but in a vestigial form and often inadequate.[40] Recently some of the smaller village ponds have been turned over to *panchayats* which are supposed to be autonomous and representative institutions of local government but are in fact neither. They depend mostly on subventions from the state governments and do not appear to take much interest to tank maintenance. In all other cases the state public works departments are wholly responsible for maintenance, the costs of which are met out of allocations in the general budget of the State governments.[41]

Users of government canals pay a water cess (which is sometimes merged with land revenue) that are supposed to cover operational costs (including maintenance and interest on capital) but are almost invariably quite inadequate for the purpose. Moreover, in contrast to China, such cesses are not earmarked for use by the system managers but get merged in the general pool of resources. Consequently, even as the PWD has to compete with other departments for allocations from this pool, the managers of different systems have to compete with one another and with claims of other activities of the department. Maintenance of irrigation systems does not appear to command a high priority in deciding budgetary allocations at the State level or within the PWD.

The complaint of inadequate allocation is one of long standing and there is evidence of the deterioration in the condition of important segments of the irrigation system.[42] And yet there has been hardly any pressure either from the PWD officials or from the users to remedy the situation. The lack of concern among officials is perhaps understandable for they are not committed to serving as managers of any particular irrigation system, nor do their careers depend on how well this job is done. The apparent passivity of users is more difficult to explain: the sheer size of the systems, the weakening of collective interest due to the development of independent, supplementary sources of irrigation (especially wells, and tube wells), and the fact that they have no role at all in the management may all be contributory.

MANAGEMENT OF WATER ALLOCATION

The Nature of the Problem

The other major task of irrigation organization is to regulate the allocation of the available water among alternative uses and between different segments of the service area. The simplest case is that of a farmer with his own irrigation source who wants to use its water along with his other resources (land, labour, equipment, and working capital) to secure the maximum net income. The available water can be used to grow various crops, the feasible set of which is determined by sundry elements that go to make up the agro-climatic environment. Within the limits set by the environment, a decision has to be made about the crops to be grown and the area to be sown with each crop. Each crop has a different water requirement, responds differently to variations in the amount of water and of other inputs. Taken together they define the 'production function' of different crops. Given this information, the price of each crop and each input, and the volume of water and other inputs available, it appears to be a fairly straightforward task to work out the optimum allocation of all inputs (including water) between crops.

In point of fact the determination of the optimum allocation even for an individual farmer wholly in command of his irrigation source is more complicated and difficult than it appears.[43] This is because, (1) soil and topography varies even within a farm; (2) the effect of irrigation on crop yields depends not only on the total quantum of water applied but also on when it is applied; (3) the response to irrigation is not independent of the quantum and the seasonal distribution of rainfall, both of which are highly variable and unpredictable; and (4) the eventual yield being the cumulative result of the moisture status at different stages of crop growth, it is difficult to evaluate the contribution of a given amount of irrigation at a given point in time without specifying the levels of irrigation at other points in the season as well as the level and quality of seed, fertilizers, and other inputs.

The problem is much more difficult in the case of multi-user systems where it is not only the allocation between crops and over time but also that between users which has to be decided. Given the inherent difficulties of evaluating the productivity of irrigation water and the fact that the supply of water and its productivity to any one user is not independent of the actions of others (e.g. upstream users) and powerful farmers can interfere with the distribution of water, it is perhaps not surprising that the market mechanism is so little used in managing water allocations in surface irrigation systems.

The alternative of using administered prices consistent with the optimum use of the available water and other resources, keeping in view specified distributional objectives, is also beset with the same difficulty.[44] There is the additional problem that irrigation systems are often designed on the basis of inadequate or even unreliable information on water supply, yield responses, irrigation efficiency, etc. 'Optimum' allocation between crops and users derived from such shaky information can scarcely be considered 'objective' or reliable guides to allocation. In any case few systems are equipped with control structures or measurement devices to permit regulation of water deliveries by volume.

Under these conditions it is difficult to decide the configuration of the 'efficient' system, and even more to ensure 'efficient' water allocation on the basis of prices alone. At any rate, most surface water systems are set up by the user community with its own resources or by the state using public funds. In either case, considerations of distribution cannot be kept out of project design or operation, both of which represent some form of compromise regarding the balance between 'efficiency' and 'equity'. The potential for conflict in the course of operation is ever present and, given the possibility, whether on account of location or power, that users can interfere with water distribution, there is necessity for physical rationing of the area irrigated (both the total and by crop) and of the amount of water supplied to particular crop in the light of available total supplies. The allocation procedures in practically all multi-user systems seek to define the basis on which these decisions are to be made and also the mechanisms and procedures by which they will be enforced.

We have descriptions of how these tasks are managed in several systems from all over Asia. Though these accounts leave much to be desired—most of them are not sufficiently detailed, often the distinction between what is supposed to be and what is gets blurred; and they reflect the sundry concerns and perspectives of the authors—they do help focus on the significant differences in practice between different types of systems across Asia.[45]

Traditional Community Systems

Descriptions of the water allocation procedure in traditional community-based systems (which represent an important, and in some countries the dominant, type of irrigation in Asia) are relatively more numerous, and some of them quite detailed. These systems vary from fairly small ones serving a single village or a part of it (like the Pul Eliya tank in Sri Lanka described by Leach (1961)) to moderate sized multi-community systems like those of Bali and Japan. They include irrigation based on storage ponds, diversion of river flow, or a combination of the two. Most of them are used essentially for paddy cultivation, which again is the dominant characteristic of irrigation systems in most parts of Southeast and East Asia, and of traditional systems in south India and Sri Lanka. They are generally relatively old systems (some are in fact several centuries old). Their current management practices are the result of a process of adaptation to changes in the land tenure system, prices, technology, and the system itself.

The water allocation problem, as we have seen, consists in delimiting the area to be irrigated and deciding the amount of water to be given to different segments of the area entitled to irrigation. The limits of the service area are more or less well defined (partly by topography and partly by the amount of water available) but variable: the area which can be effectively irrigated in any given season depends on the supply available in the storage or river flow. Moreover, the area sought to be irrigated as well as the supply of water may show secular change: the service area may be extended (on account of increased demographic pressure or as a matter of policy), and the effective supply of water may change for better (by extending and/or improving the system), or for worse (on account of negligent maintenance, damages from war or natural events and, possibly, long term climatic changes).

The extent of area to be cultivated as well as the timing of the start of irrigation in a particular season appears to be generally decided by the irrigation community as a whole in the light of available water supply and rainfall at the beginning of the season. Where the entire service area cannot be irrigated, there is the problem of deciding on the basis of rationing supplies. This appears to vary widely. Some systems have worked out quite elaborate and ingenious arrangements to ensure equitable sharing of the shortfall in supplies among all users: for instance in the case of the Pul Eliya tank (Sri Lanka) the irrigation area is divided into three segments according to distance from the tank and the wetland holdings of every farmer are equally supplied between these parts. Because of this:

If the villagers are to cultivate rice in the old field during the Yala season, they will decide from the start either to cultivate the whole of the field or the upper two thirds of the field or just (the uppermost) one third.... No pooling of proceeds or reallocation of holdings is necessary since the land is already divided up in such a way that each landholder works the whole or two thirds or one third of his total holdings as the case may be.... [Leach, 1961.]

Chambers (1977) cites instances from south Indian tanks where the rationing of area is done on different criteria. In one case each user was guaranteed full irrigation for a limited acreage; any additional area being given water only if there was any surplus. In another, the farmers were left free to decide how much they would sow. Thus rationing is not always on the basis of a proportionate reduction in the irrigated acreage of all users.[46]

In many systems (e.g. the Balinese *subak*, the Japanese systems, as well as in Thailand and Taiwan) it is an established custom that some sections of the service area have superior rights and hence a first claim over available supplies.[47] In general, this differentiation is based on location: upstream lands and communities have priority over those downstream. While this is partly a formal recognition of the inherent advantages of upstream users, it often happens that they were also among the first to develop the irrigation source.

Acreage rationing is but one element of the water rationing system.

There remains the problem of regulating the quantum of actual supply in accordance with the needs of crops in different sections of the command area in the course of each season. Even in a normal season, these systems have both physical devices and operational procedures (that can be quite elaborate and ingenious) to divide the available supplies among different plots on some recognized basis.[48] In times of shortage, besides rationing acreage, the water allocation is regulated even more strictly.

The basis and procedures vary. Sometimes (as in Pul Eliya) the physical design of the system is such that it is possible to ensure that the reduction in both area and water supply is shared in more or less the same proportion by all users. This however appear exceptional: the area of each landholding is not usually distributed so systematically across lands with different grades of water rights. In such cases, only rationing between different sections of the command area is attempted. Rotation by canals, distributaries, outlets, or a combination of these appears common. Not infrequently this is subject to differential water rights of different segments of the command area. There are a few cases where the supply of water for relatively water intensive crops is rotated from year to year: this is usually in areas where crops other than paddy or more than one crop of paddy is raised in a year and the available supply of water even in a normal year is insufficient to permit all users to grow the water intensive and dry season crops.[49]

Coordination of irrigation schedules with crop water needs is facilitated in some cases by collective agreement on the schedule of the cultivation operation in each season. Thus, in Pul Eliya, traditionally, the villagers take a collective decision at the beginning of each season regarding which land and how much is to be cultivated as well as the schedule regarding 'the dates on which the sluices will be open, the date at which sowing will be completed, the varieties of rice which will be sown and the date by which it is planned to have the harvest ready' (Leach, 1961: 108).[50] In some Japanese systems the fact that the date of releasing water for puddling and transplantation is decided by the irrigation community as a whole implies a certain coordination in the timing of operations.

Where the system is relatively extensive and the seasonal distribution of water supply does not permit synchronized operations over the entire command area, cultivation is staggered. Sometimes, as in Bali, this staggering is elaborate and carefully coordinated in an institutionalized way (see Geertz, 1967). Staggering operations of a less controlled character is reported in several cases.[51] There are few systems (e.g. some of the Japanese pond-cum-canal systems and the Meichuan project of China) that operate flexible delivery schedules which are varied according to the crop water needs in different segments during the crop season.

The individuals or groups responsible for the overall management of the system make the allocation decisions within the framework of principles established by convention (or, in, some cases, like the Sri Lankan tanks, by state law) and supervise its implementation. In multi-community systems the function has to be distributed between different layers, each layer being responsible for allocation within its jurisdiction subject to the limits set by the higher level. The basic unit at the field level is usually a group of farmers served by the same outlet. While users and their representatives participate at all levels, knowledge, experience, and skills in water management are given considerable weight in the choice of functionaries.[52] In larger multi-community systems, where a higher order of expertise is necessary, specialists are hired by the organization. In this way technical expertise and users' interests are combined in arriving at decisions.

Where the system is fed by single pond or barrage, the top level decisions in canal operation largely determine the amount and timing of supplies at lower levels. The principal task of the latter is to ensure that the supplies made available at that level are put to effective use and distributed 'fairly' among its constituents. Systems with multiple sources and/or intermediate storages permit greater flexibility of allocation at all levels, but the managerial task is much more demanding in terms of the volume and quality of information required, alertness of the operating personnel, speed of communications, and coordination of decisions. All this also requires a high degree of rapport and exchange of information between operators and users.[53]

Policing the system, which is essential to ensure enforcement of the water allocation, especially in times of shortage, is provided for at all levels through a system of guards and watchmen who are either appointed as regular paid employees (this practice being usually restricted to the higher levels of relatively large systems) or by assigning some of the farmers (sometimes for long periods and sometimes by rotation) to do this duty. The latter practice is common at the lowest level of multi-community systems.[54] In either case, guards are selected for their reliability and fairness, and remunerated by the members of the group either directly or through a levy paid to the central organization.

Conflicts occur both between constituent units of the system and among members of each unit over the way allocations are managed and over attempts to violate the allocations. There appear to be well-established conventions as to how such situations are to be handled. The leaders of a particular constituent group are expected to mediate disputes among its members and at the same time to represent the interests of their members as a whole in relation to other units of the system. Disputes that cannot be resolved at one level are referred to the next higher level. Conflicts which cannot be resolved by the irrigation organization may have to be referred to higher centres of power. It does not however follow that the latter will always intervene decisively. Unfortunately these are only few detailed accounts of the nature and incidence of actual conflicts over water allocations in such systems or about how they are resolved. The general impression that one gathers is that while conflicts are quite common and sometimes violent, there are countervailing forces that keep the organization from breaking down.[55]

Persistent conflicts that cannot be resolved satisfactorily within the existing framework is one of the factors that could trigger modifications in the organizational framework and, failing that, a redesign of the system itself. This is illustrated by the experience of several projects in Japan (e.g. the Aka river system, Azusa river, 12 Go Canal), China (the Meichuan project), and Taiwan (Nan Hung).

These cases however show that the intensification of conflicts in a system is only a signal that changes in its operational procedures and perhaps physical design are necessary. Whether and when the changes, and in particular system redesign, in fact take place depends on the potential returns to the investment on system improvement as against other means of raising output; the ability to mobilize resources to finance such investment; and on the manner in which the costs and benefits from modification will be distributed.

Thus in Japan major system improvements took place in stages spread over several decades in spite of the fact that these improvements meant more abundant and assured water supply and hence a significant reduction in conflicts among users. Much the same was the case in the Nan Hung project of Taiwan. In Meichuan, on the other hand, such improvements appear to have been brought about in a remarkably short period and the necessary resources mobilized wholly from the irrigation community because the combination of the commune system and strong local political leadership made it possible to secure an enforceable solution. (Nickum (ed.), 1981.)

That a high level of conflict or increasing conflict does not necessarily lead to such improvements is highlighted by the state of tank irrigation systems of south India and Sri Lanka. In both these regions, irrigation supplies in relation to needs are much scarcer than in East Asia, and given the low and variable rainfall in relation to crop water needs, one would expect the potential returns to irrigation to be greater. It is true that the supply of water in these systems is much less reliable, and therefore the returns are more uncertain, than say in Japan. Also, the process of breeding superior crop strains and improving cultivation practices started much earlier in Japan. Major changes in biochemical technology have been slower to develop in South Asia. This does not however appear to provide an adequate explanation for the continued neglect of tank maintenance and the inability to contain encroachment of cultivation into tank beds and other 'prohibited' areas, which have reportedly led to a progressive reduction in storage capacity and hence water supply.

In more recent times, despite the rising prices of paddy and the introduction of HYV, no major changes appear to have occurred in the management of tank irrigation in South Asia. The organization and leadership needed to convince users of the necessity for such changes and allay their legitimate fears, and to convince them that their interests will be protected—both steps entailing a sustained process of education, persuasion, and pressure—does not exist and has been slow to develop. These difficulties are to some degree being sidestepped by recourse to groundwater pumping on an individual basis, a phenomenon that has by all accounts become widespread even in tank irrigated areas.[56] However, there are hardly any studies of irrigation from this perspective to enable me speak with confidence on this aspect.

Large State-Managed Systems

At the other extreme, and in sharp contrast to the above category of systems, are the large canal irrigation projects of India (and South Asia). The management of water allocation in the latter is conditioned by the following characteristics: they serve extensive areas; until recent times, most of them were fed by a single source (usually a reservoir); in contrast to most parts of Southeast and East Asia, they cater to a wide variety of crops with different growing seasons and growth periods; and the allocations are made by the system managers without consulting the users. In large systems with numerous users, communication is more difficult, changes in any particular segment could have ramifications over a wide area, and the task of coordination becomes more complex. In consequence there is a preference for relatively simple operational principles which necessarily involve a certain rigidity in schedules and procedures. The scope for flexibility is further reduced when the system receives all or the bulk of its supply from a single barrage or reservoir. The diversity of irrigated crops makes coordination of water deliveries with crop water needs more complex than in systems that utilize irrigation largely for paddy.

In general, the Indian canal system[57] is geared to a specified crop pattern (in terms of area under paddy and other seasonal crops and perennials by major crop season). This is supposed to be decided and incorporated into project design on the basis of (a) a careful evaluation of the quantum and seasonal distribution of rainfall; (b) likely water supply in the reservoir; the water requirements of different crops under different soil conditions; and (c) the policy regarding how the benefits are to be distributed. The extent of area that can be sown with certain water intensive crops (like paddy and sugar-cane) is also specified with provision for penalties (penal water rates and even denial of water) in the case of violations. In some cases, annual rotation of supplies between different distributaries fed by a main canal is incorporated into the design to prevent excessive concentration of water intensive crops in the command area. In general the system is designed for continuous operations of the main canals except during the closure period which varies, depending on the needs of maintenance and on the crop pattern. The operation of branch canals and distributaries are continuous in some systems, rotated in others.

There are, however, important differences in the way the canal supplies are regulated in any particular year or season. In south India, depending on the supply in the reservoir, water is supposed to be released to the command area from a fixed date at the beginning of the crop season till the maturity of the main crop:

... the main canal runs continuously and the distributaries may be rotated in some projects with limited control structures. In this system no effort is made to ascertain the demand of the farmers on a weekly or fortnightly basis which vary due to rainfall and the staggering of crop sowing from distributary to distributary and year to year. [Kathpalia, 1980: 6.]

The focus of equitable distribution in south Indian system appears to be more on the regulation of crop pattern, particularly with regard to water intensive crops, both overall and in different segments of the command area.[58]

The annual operation of the systems in west India is also decided on the basis of the supply position in the reservoir, but deliveries in different branches and distributaries are regulated on the basis of the area under different crops (especially the water intensive crops) in different parts of the command area sanctioned by the canal management on application by farmers at the beginning of each season. In these regions, the water supply in the winter season is rotated by distributaries as well as within each distributary.

In north Indian systems, while 'the supply of water to different distributaries and minors is worked out in advance depending on the forecast of available water every season,' the procedures are meant to give 'flexibility to adjust the supply in the course of the season' (Kathpalia, 1980). This is done on the basis of 'demands' received from farmers, screened by the successive levels of canal officials, and channelled up to the authority controlling releases in the main canal. This process is supposed to be done at frequent intervals during each season, for each minor and distributary, taking into account 'the rainfall in the area, the stage of crop growth and the area under different crops at the particular time,' care being taken '... to limit these demands within the available water in the particular crop season of the year' (ibid.).

Rotation of supply, usually by multiples of a week, between branches, minors, and distributaries is common, especially during the winter season. There is reportedly no rotation between outlets located on a given distributary. The procedure of the north Indian systems would appear to have a built-in method of handling situations of shortage. The fact that water is supposed to be released on the basis of the farmers' demand subject to official review in the light of overall water supply provides a potential means for enforcing an 'equitable' rationing of water in periods of shortage. Other systems do not seem to have any well-defined procedure for handling shortages. The reactions of the canal management seem to be more or less ad hoc.

In point of fact, however, even in north Indian systems, scheduling is much less flexible than the above description would suggest. A detailed study of the working of the allocation procedures in the Bhakra canal system showed that (in the late sixties), while the rotation schedules as between the distributary channels enabled farmers served by each channel to be reasonably certain as to the date when they might expect to get supplies, it gave no guarantee as to the amounts they would get. The latter depended on the amount of water released in the system which was highly variable. The uncertainty of timing and volume of supplies at the individual farmer level, which is determined by a further rotation system below the channel outlets, is considerably greater (Reidinger, 1974).

Where rotational scheduling is not in vogue, as is typically the case in south India, and there are no codified procedures for managing drought situations, there is not even the pretense of attempting to assure predictability of supplies and equitable distribution of water to each outlet. (The Parambikulum Aliyar Project in Tamil Nadu is a particularly striking instance, see Vaidyanathan and Janakarajan, 1988). In effect, the area close to the head of the distribution channels have greater assurance of supply, the degree of certainty falls sharply as one moves towards the tail end. The experience of one canal system during a recent drought highlights how slow the management is in reacting to water shortage, the ad hoc manner in which it attempts to cope with the situation, and also the kind of pressures under which it has to operate (Wade, 1980).

More generally, few systems appear to be able to conform to allocations visualized in the original design. The planned allocations are based on assumptions regarding, (a) the likely availability of water in the system, its seasonal distribution, and variability; (b) the losses in conveyance, distribution, and field application of water which together determine the 'technical efficiency' of the system; (c) the water requirements of different crops; and (d) the pattern of cropping in different sections of the command area. Many of these turn out to be either erroneous or mutually inconsistent.

The available information on rainfall, stream flow, and crop water needs are often inadequate or unreliable. With the best of intentions, the project design calculation can and often do go wrong.[59] In the post-Independence period, inadequency of time and staff devoted to surveys and design in relation to the volume of work involved, along with laxity in the technical and economic scrutiny of projects before approval, have contributed to a significant lowering in the quality of designs. That there is political pressure to make it appear that project benefits are to be widely distributed is an aggravating factor. Under these circumstances it is difficult to ensure that the availability of water, the size of irrigable area, crop pattern, and the water allocation plan are at least mutually consistent, not to speak of 'efficient' or 'optimal'.

Such defects in design are compounded by difficulties in enforcing the planned crop patterns (in terms of both extent of area under different crops and their location). Most Indian canal systems are characterized by a shortage of water in relation to irrigable land. The rainfall in relation to crop water needs is typically low over the larger part of the year. Consequently, the potential returns to irrigation by way of increased yield of particular crops and, even more, through shifts to high value crops, is large. This offers a powerful incentive to violate both the crop pattern and the irrigation quotas.

The incentives are not neutralized by appropriate differential pricing of water by use. The physical control devices are too crude to permit effective regulation of the volume of water delivered to particular locations necessary for proper rationing of water supply. The canal bureaucracy has in principle a variety of ways of enforcing the desired crop patterns. Some of these (like localization, regulation of canal supplies, and policing) have to do with operational procedures, and others involve invoking sanctions against violators. They do not however appear to have been used to much effect.[60]

Since the available supplies are inadequate to meet the needs of the entire service area, and since the timing of supplies is highly uncertain, one naturally expects conflicts over water allocation to be relatively intense and widespread in the Indian canal systems. Conflicts among different segments of the command (up to the outlet level) are supposed to be mediated by the canal bureaucracy. There is no mechanism in the formal framework by which users interests and operational constraints can be made to confront each other and compromises evolved. When authorities fail, or users are disaffected, extra-institutional channels of influence and pressure are invoked.

Reduction in the area effectively irrigated is one way by which the conflicts over water allocation appear to be contained. This happens not so much by design as by the head-reachers exploiting their advantageous location. There is one instance where the canal managers sought to invoke penalties over such violations but gave up because the latter could muster the necessary support at higher levels of government to thwart the enforcement of regulations.[61] It is somewhat surprising that this does not appear to have led those at the periphery and the tail end to mount a sustained and organized effort to seek redress.[62]

The reduction in irrigated area does not of course eliminate 'conflicts' over water. Again, not much is known about their nature and magnitude, where they occur, how they manifest themselves, and how they get resolved. All we know, in a general way, is that water is diverted from canals and channels in violation of rules, sometimes with the connivance of officials, sometimes due to their negligence and, on occasion, despite their efforts to enforce the regulations. Apart from the location of the channels (as we have noted, upstream channels can get away with 'violation' more easily than tail enders), the ability of a group to influence the officials through persuasion, for a consideration or through 'political' pressure, is quite important.

We know even less about the way allocations below the outlet are managed. In north India, a system of rotational supply in accordance with cropped area to be managed by users, but backed by legislative and administrative support, was introduced in the late nineteenth century and appears to have become a general practice in all systems. In other parts of the country they have been slow to develop. The field channels are not constructed, and the distribution below the outlet appears anarchic.

During the past decade or so, a conscious effort has been made, as part of the Command Area Development Programmes, to improve the field distribution system both in its physical and organizational aspects. The introduction of a rotational system within the command area of each outlet in all parts of the country is receiving serious attention. This has largely been at the initiative of the state, though more spontaneous forms of community effort have also been reported (Hashim Ali, 1980; Jayaraman, 1981; Singh, 1983). It has been suggested that the allocation at this level are biassed in favour of the larger and more powerful landowners.[63]

However, a detailed study of one outlet in the Bhakra system (Vander Velde, 1970) suggests that when water available at an outlet is inadequate in relation to its command area, rationing appears to take place essentially on the basis of the location of plots. Those closer to the outlet receive more water and more assured water than those at the tail end. However, the present state of knowledge on the management of allocations and conflicts at the field level is too rudimentary to permit any generalization.

In all these respects the situation is far from static. Significant changes have taken place both in the organization and techniques of water management. I have already referred to the example of the Sarada canal where the introduction of the government canal appears to have supplanted pre-existing local systems and the institutions associated with them. The canal systems of the Deccan, which were originally designed for protective irrigation but then altered to support sugar-cane cultivation, is another case in point. The consolidation of landholdings in north-west India (especially west Uttar Pradesh, Haryana, and Punjab), and the abolition of zamindari may also have induced some changes, but there are scarcely any studies documenting them. Several changes in both the organization and in techniques of water management are currently being instituted.

In recent times awareness of the need for improved water control has increased, possibly as a result of the spread of HYVs (whose high potential cannot be realized without adequate and timely supply of water). To some extent the high and rising costs of new irrigation development and the fact that in some areas the scope for extending the irrigated area is reaching the limit, have perhaps helped to shift attention towards more effective use of available supplies. The encouragement of ground water use in the command area of canals, which was earlier prohibited, is a major change. It has enabled the effective supply of water in the canal irrigated areas to be increased and at the same time given farmers greater control over timing and volume of application. The other important development is the establishment of the Command Area Development Programmes aimed at increasing the efficiency of water use by improving the physical facilities, planning the crop pattern and water deliveries on the basis of more careful studies of soil conditions, irrigation efficiency and crop water need, and introduction of new systems of water distribution both between different parts of the canal system and among farmers served by a particular outlet.[64]

These Programmes have not accomplished all that they set out to do. Major changes in the basis of regulating distribution of canal water has proved difficult even where they have been shown to facilitate wider sharing of available supplies without reducing yields (see Wade, 1980). Attempts to bring about improvements at the user level by the construction of field channels, consolidation of holdings, and land levelling have not made much headway. The uncertainty of returns to the required improvements which involve substantial investment, the fact that the efforts are almost wholly conceived and implemented by state agencies with little user involvement, and the organizational weakness of the state apparatus are all important constraints.

Given these constraints, it is perhaps not very surprising that the response to higher returns to irrigated agriculture has taken the form of expanding conjunctive use of groundwater with canal supplies. The necessary investment for this purpose is made mostly by the better-off farmers with the smaller, less wealthy cultivators buying water from the former. While the distributional consequences of these changes are a matter of legitimate concern, the relevant point in the context of the present discussion is that they provide evidence of adaptation to changing circumstances.

CONCLUSIONS

The central argument of this chapter is that a proper understanding of the form and effectiveness of irrigation organizations requires that their structure and functioning be viewed in the context of agro-climatic, technological, and socio-economic environment in which they operate.

I have suggested, with some illustrative comparisons between China, India, and Japan, that the nature of water control necessary for efficient agriculture as well as the kinds of works that are feasible in a given situation are strongly conditioned by the level, seasonal profile, and predictability of rainfall as well as by temperature, geology, topography, and other features of the physical environment. Within the limits set by these physical and environmental factors, the scale of water control development as well as the nature, size, and sophistication of the sysems are influenced by the state of technique both in hydraulic engineering and in agriculture, the relative costs of raising production through different means (including different types of water control); and, not the least, the interests of the various parties involved in the process.

The nature and size of systems has a bearing on the way their construction and management is organized. While groundwater is mostly developed by individual farmers for their own use, surface systems under Asian conditions invariably serve several users and are subject to some form of collective control. Small systems are generally constructed with local resources and leadership. However, larger systems tend to attract the involvement of the state or, to be more precise, supra-local authority. Such involvement is necessitated by the higher order of design and engineering skills required to construct large systems and also to decide the limits of the command area and the basis for allocation of water among potential beneficiaries. That state involvement even in large projects does not necessarily imply state financing of their cost is evident from the Chinese experience.

Small systems everywhere tend to be managed informally with scarcely any bureaucracy. The degree of formality and reliance on bureaucracy necessarily increases with the size of the system. But whereas in East Asia, users have been actively involved in setting up a system, a high degree of user control on management is noticed even in moderately large systems. In sharp contrast, the canal systems in India are managed by a highly centralized bureaucracy with little or no participation, not to mention control, by users.

A striking feature of both community and state managed surface systems is their reliance on physical rationing of supplies rather than the market mechanism or prices to regulate water allocation between users and uses. The market mechanism is scarcely in evidence among surface systems, at any rate in Asia. Administered pricing is common, but invariably combined with regulations concerning crop patterns and water deliveries. Small, relatively old systems do have accepted conventions concerning water allocation both in normal times and in times of scarcity. In new systems, particularly those constructed wholly by the state, the allocation postulated in the project design may not be acceptable to all segments of the users, and usually be modified once the system begins to operate. Conflicts are pervasive in terms of scarcity and even in normal times, when there are large differences between the outputs which unrestricted access can produce, and what is feasible if the stipulated allocation is strictly observed.

Irrigation organization needs to be viewed from an evolutionary perspective. The way a system is designed in effect reflects the state of knowledge, techniques, skills, and prices, as well as the interests of the various groups involved at that time (the latter including not only potential users of the system but also those who organize and finance the construction). Since designs are based on limited and often erroneous understanding and facts, the planned allocations are at best approximate and, not infrequently, defective. The effective command area, the crop pattern, as well as the operational rules get modified in actuality, even if not always formally, in the light of actual experience regarding water availability and its seasonal distribution, crop water needs, and the effects of varying water (under the prevailing agricultural techniques) on crop yields until a certain 'equilibrium' is established. At this stage any hiatus between the system authority's objectives and the users interests have also to be sorted out, whether by consensus or fiat or by users in the upper reaches asserting their locational advantage.

This position can be disturbed for any of a number of reasons: other groups may begin to tap the same source (by extending existing canals or by taking out new ones from the same river), thus leading sooner or later to shortages and conflicts. Neglecting to maintain system facilities may reduce effective supplies and aggravate conflicts. Expansion of the market crops, introduction of new crop species and varieties, improvements in cultivation technique (including new techniques of irrigation), a shift in the prices of crops in relation to inputs may all individually and collectively increase the potential returns to irrigated agriculture. Changes in land tenure and taxation may alter the returns which the actual cultivators get from irrigated farming. If there is no change in the system or allocation procedures, these changes, insofar as they increase the potential returns to water, could aggravate conflicts over allocation.

It would appear natural that when conflicts over water in a given system increase, whether as a result of over-extension of the command area in relation to supplies or because water is becoming more valuable, attempts will be made to change operational procedures and to make minor modifications in the physical facilities with a view to reducing waste and facilitate a more equitable sharing of available supplies. This process however has definite limits beyond which further improvements in the technical efficiency of irrigation and in the regulation of the volume and timing of irrigation requires major modifications in system design (involving such works as integration of intakes, increasing water supply by constructing new storages and/or tapping groundwater, redesign of canal and drainage networks, and land improvement) or introduction of improvements (such as land consolidation and levelling, better techniques of water application) which raise irrigation efficiency. All of this entail additional investments. The willingness of beneficiaries to undertake these investments or to contribute to the cost depends as much on their overall productivity in relation to the costs involved as on their perception of how the changes in system and its operational rules will affect their individual interests.

Insofar as changes in organization and its operational procedures are essential to get the most out of these system changes (in terms of additional output), the two cannot be dissociated. At this point again a difficult and time-consuming process of securing, through persuasion or pressure, the users to accept the organizational changes is invariably involved. Even as procedural changes cannot go beyond a point without system modification, institutional changes must sooner or later accompany system changes.

The few studies of the evolution of irrigation systems and their organization which we have broadly concur with the above view. Most of them however relate to the relatively small irrigation systems of East Asia which have grown by combining, expanding, and improving pre-existing local systems. The pace and character of these changes however vary a great deal, and not much can be said with confidence on the reasons for these variations. Besides, there are hardly any studies of the changes that have occurred, and are occurring, in the very different and much more varied, irrigation systems of South Asia: (a) the reasons for the apparent failure of traditional tank irrigation systems to respond to changes in cultivation techniques and prices; (b) the difference in response between users (in terms of community organizations to maintain local facilities and allocate water within the community), to introduction of canal irrigation in areas that already had local systems, and those which had no prior tradition of irrigation; (c) the manner and speed with which users in different segments of a large canal system respond organizationally to deal with the allocation problem; (d) the differences in response between areas with groundwater and those without; as well as (e) between areas experiencing high rate of technical improvement and those which have experienced little appear to be some of the issues that seem promising subjects for systematic investigation.

ACKNOWLEDGEMENTS

Much of the work on this chapter and that on China was done during my tenure, in 1982, as visiting Research Fellow at the Institute of Developing Economies, Tokyo. Besides giving me an opportunity to gather material on East Asian irrigation systems, the Fellowship enabled me to discuss the subject at a more general level with several Japanese scholars, most notably the late Professor Tamaki, Prof. I. Hatate, Prof. H. Okhomoto, Prof. S. Ishikawa, Dr R. Kojima, and Mr H. Nakamura. I am grateful to all of them. I have also benefited from critical reactions to earlier drafts from colleagues at the Centre for Development Studies and the Madras Institute of Development Studies, and from the referee's comments on revising the original paper on which it is based.

NOTES

1. Marx suggested that 'climate and territorial conditions' made artificial irrigation by canals and waterworks the basis of oriental agriculture, and 'this prime necessity of an economical and common use of water which, in the occident drove private enterprise to voluntary association as in Flanders and Italy, necessitated in the orient where civilization was too low and the territorial extent too vast to call into life voluntary association, the interference of the centralising power of the Government' (Marx, 1853 cited by Wittfogel, 1957 and Chi, 1936). In relation to China, Chi interpreted the comment about the 'low level of civilization in the orient' in terms of the tendency of the ruling 'landlord bureaucracy' to nip the growth of merchant capital in the bud by declaring 'every important profitable enterprise as a state monopoly and absorbed the embryonic merchant class into their ranks if it were felt it would be unsafe to disregard it' (Chi, 1936: 71).

2. Weber argued '... in the cultural evolution of Egypt, West Asia, India, and China, the question of irrigation was crucial. The water question conditioned the existence of the bureaucracy, the compulsory service of the dependent classes, and the dependence of the subject classes upon the functioning of the bureaucracy of the King' (Weber, 1927 cited by Chi, 1936: 73).

3. His ideas are set out in a series of works, the first of which appeared in 1926 and culminated in the book entitled *Oriental Depotism: A Study of Total Power* (1957). The latter volume provided a resume of the evolution of Wittfogel's ideas on the subject and also their antecedents.

4. Leach (1961) formulated the perspective thus: The material context of societies '... is not merely a passive backcloth to life; the context itself is a social product and is itself structured. The people who live in it must conform to a wide range of rules and limitations in order to survive at all. Every anthropologist needs to start out by considering just how much of the culture with which he is faced can most readily be understood as a direct adaptation to the environmental context including that part of the context which is man made' (Leach, 1961: 306). The perspective is however not universally accepted and, as Leach goes on to point out, 'While every social anthropologist recognized that societies exist within a material context... too many authors treat such things as nothing more than a context useful only for an introductory chapter' (Ibid.).

5. Of these, Kelly's study of the evolution of the Aka river system in north Japan spanning a period of nearly three centuries (from its inception in 1600 to 1870) is perhaps the most detailed. For a more general treatment of the history of irrigation in Japan, and its relation to evolution of land tenure, see Hatate (1978).

6. For a useful and non-technical summary of the present state of knowledge on the role of water in plant growth see Dakshinamurti et al. (1973) and Carruthers and Clark (1980, Ch. 2).

7. The determinants of ET and its relationship to crop water needs are of course more complex than this. There are also many controversial and unresolved

questions. However, for our purpose it is appropriate to focus on two basic propositions that are more or less generally accepted: (1) the maximum crop water needs are a function of evapo-transpiration; and (2) they do not vary much over a wide range of field crops. For an idea of the nature of other influences on crop water needs see Weisner (1970); Carruthers and Clark (1980); and Dakshinamurthi et al. (1973). For a discussion on the determinants of ET and the problems of measurement see Weisner (1970) and Olivier (1961).

8. In the subsequent discussion I do not always distinguish paddy from other crops. Paddy is unique among field crops, in that it needs substantial amounts of water over and above ET both for keeping the field submerged and for puddling and transplantation. Percolation losses from paddy fields also tend to be relatively high. This may not pose a serious difficulty when comparisons are made between countries/systems where paddy is the dominant crop. However, in comparing systems/regions where paddy is grown along with other crops in irrigated tracts, and this is quite common in South Asia, these differences in water needs of paddy and other crops become important and need to be explicitly taken into account.

9. The pattern in peninsular India is broadly similar to the pattern shown in Fig. I. However the rainfall, even during the monsoon, is inadequate to sustain paddy cultivation whose water requirements, for reasons cited, far exceed the ET. In 'arid' regions (like Rajasthan and north-west India) rainfall is uniformly below ET.

10. This account of the principal features of the organization of water-conservancy planning and construction in India is based on a variety of sources including the various five year plans and the report of the Irrigation Commission (1972).

11. This description draws heavily on Nishimura (1971); Vermeer (1977); and Greer (1979). See also Ch. 4 for more details.

12. A number of descriptions of the post-war land improvement programmes are available. See, for instance, Ogura et al. (1963: Chs 12 and 20), Nishikawa (1971).

13. Saradaraju (1941: 113–20) gives a brief review of the situation immediately before the institution of British rule and thereafter up to the end of the nineteenth century.

14. For a recent historical account see Whitcombe (1983).

15. For more detailed accounts see Ch. 4.

16. For a description of the historical evolution of particular Japanese irrigation systems and the role of government in it, see Tamaki (1979), Hatate (1981), Kelly (1980), Shimpo (1978).

17. Out of a total irrigated area of 43.3 m. ha., 26.7 m. ha. were served by surface sources consisting of canals (19 m. ha.) tanks (4.1 m. ha.), and other surface sources (3.5 m. ha.) (Rao, 1979: 59). Canals thus account for nearly 70 per cent of the total area under surface irrigation.

18. Thus the effective storage capacity rose from an estimated 12.3 bn. cu. m in 1951 to 61.7 bn. cu. m in 1966 and 162 bn. cu. m in 1990. (Framji and Mahajan, 1969, Vol. 1: 420, GOI, Central Water Commission, 1992).

19. In the mid-fifties, there were some 41 multipurpose storage dams with a total storage capacity of 1.4 bn. cu. m (bn. = 10^9) and 50,000 small and medium reservoirs with an aggregate capacity of 1.64 bn. cu. m (Sasaki, 1959: 13–4).

20. In the early 1970s there were some 10,000 land improvement districts irrigating 1.4 m. ha. The average LID thus covered about 150 ha. About a third of them served less than 50 ha. only 2 per cent served over 3000 ha. There were only 17 systems serving over 10,000 ha. (Takeuchi, 1979: 87).

21. At the time of Independence most of the canal irrigation in the Indo-Gangetic plain consisted of diversion works. Among the prominent ones are the Sirhind canal (0.6 m. ha.), the Upper Bari *doab* (0.33 m. ha.), the Son canal (1.35 m. ha.); the Jamuna canals (0.68 m. ha.), and the upper Ganga canal (0.7 m. ha.) (Rao, 1979: 241). It should be noted however that the overall level of irrigation development was low and the bulk of irrigated area was under small local works and wells. See Sengupta (1980) for a description of some of the local systems in south Bihar and Whitcombe (1972) for a description of their role in Uttar Pradesh.

22. The largest of which is the Bhakra Nangal reservoir (capacity 7.4 bn. cu. m). As of 1979, the total storage capacity in the Indus and Ganges basins (including multipurpose reservoirs) is estimated at 48 bn. cu. m representing nearly a third of the total storage capacity of all reservoirs in the country (147 bn. cu. m). Practically all the reservoirs in the Indo-Gangetic basin were constructed since Independence (Rao, 1979: 239–40).

23. Nearly half the irrigated area in the Ganges basin (8.9 m. ha. out of 19.5 m. ha.) and about two-fifths in the Indus basin (2.4 m. ha. out of 6.3 m. ha.) is estimated to be served by ground water. More than half the area irrigated by groundwater sources is currently under tubewells which is entirely a post-Independence phenomenon (Rao, 1979: 59).

24. There are an estimased 1,27,000 tanks in the area covered by the present states of Andhra Pradesh, Karnataka, and Tamil Nadu irrigating an estimated 1.6 m. ha. (GOI, COPP,1759). Besides topography, the fact that these tracts have a raintall pattern marked by two distinct seasonal peaks (one during May–July and the other in October–December) is an important factor facilitating tank irrigation. For a detailed study of tank irrigation in south India see Vaidyanathan (ed.), 1998.

25. In the river basins of south India (Godavari, Krishna, Cauvery, and Pennar) only 25 per cent of the irrigted area is served by wells, and that too mostly shallow ones. Of the area served by surface water, over one-fifth is under tanks; the comparable proportion for the rest of the country being 10 per cent (Rao, 1979).

26. There is obviously great variation in the climatic patterns within east Asia. My interest here however is to focus on the gross differences between east and South Asia. These differences appear to me to be very striking and have an important bearing on the nature of the water control problem.

27. See Maass and Anderson (1978, Ch. I) Maass, Coward Jr. (1980: Ch. I), Chambers (1977) for a generalized discussion on functions of irrigation organizations and of the conceptual framework appropriate for studying their structure and working.'

28. In Japan, 10 per cent of the irrigated area is reported to be under individual management (Takeuchi, nd: 87). These systems are presumably also owned by individuals. In India, a survey conducted in the 1950s reported 30 per cent of the tanks as being privately owned, a large majority of them jointly (GOI, PEO, 1961).

29. See Sasaki (1959).

30. For details see Ch. 4.

31. For a general description of organizational structure see Framji and Mahajan (1969) and Govt. of Andhra Pradesh (1982: 78–87); GOI, PEO (1965) gives some, albeit sketchy, ideas regarding management organization for selected projects.

32. For a discussion of the relation between water users and water authority in community systems see Coward Jr. (1980: 25–7). For a description of the characteristics of 'managers' see Potter (1971), Beardsley et al. (1959), and Lewis (1971).

33. For details see Ch. 4.

34. For descriptions regarding maintenance operations in concrete situations see Potter (1971), Myers (1975), Kelly (1980), Nickum (ed.) (1981), Lewis (1971), Bacadayan (1973).

35. For a discussion of the changing role of landlords in the Japanese context see Hatate (1978).

36. See in this context Leach (1961), Roberts (1967).

37. This is evident from the data and observations presented in GOI, COPP (1959, 1960), and GOI, PEO (1961, 1965).

38. This practice is reported in the Twelve-go canal of Japan, the Balinese Subak as well as in some Philippine systems.

39. We have an account of the lower Yangtze delta where the maintenance system broke down as a result of disputes over sharing of maintenance costs following a change in the pattern of landownership, and the state was unable to resolve it satisfactorily for several decades. Though the account pertains to the sixteenth and seventeenth centuries, it does highlight the interrelation between water control maintenance, land tenure, and bureaucracy, as well as the interplay of interests at different levels (Hamashima, 1980).

40. For further details see Ch. 4. Subsequent studies however suggest that community participation in tank management, at least in south India, is more widespread than is commonly supposed. See Vaidyanathan (ed.) (1998).

41. For a more detailed account see GOI, PEO (1961: 98).

42. See for example, GOI, PEO (1965); the papers presented at the 1980 seminar on *warabandi* held at Hyderabad; and Singh (ed.) (1982). The problem is also discussed in more general terms by GOI, Irrigation Commission (1972), and Govt. of Andhra Pradesh (1982).

43. For a discussion of the complexity of water productivity relations see Minhas et al. (1974).

44. Some attempts at modelling optimum water allocations have been made in the Indian Context. See for example Minhas et al. (1974) and GOI (1970).

45. Similar studies are available for other parts of the world, e.g. Glick (1970). Hunt and Hunt (1974), and Maass and Anderson (1978). I have not made any attempt to bring them into my discussion largely because they relate to altogether different agro-climatic, technological, and socio-economic environments that are not comparable to Asia. The dominance of paddy and small-scale farming are two distinctive features that the irrigation communities of Asia have in common, and which marks them out from those of Europe and America.

46. For information based on more recent research on this aspect, see Vaidyanathan (ed.) (1998).

47. The tradition of unequal water rights is well documented in the case of Japanese and Taiwanese systems. See Beardsley et al. (1959); Hatate (1981); Shimpo (1978); Vander Meer (1977). Similar instances have been reported in India: See Jayaraman (1981), GOI, COPP (1961).

48. For descriptions of such devices see Leach (1961), Kelly (1980), and Beardsley et al. (1959). Ingenious as these devices may be, they are not always based on any careful calculation. By modern standards they are often crude, many of them were set up long ago, and few people now can even recall the basis on which they were designed, and yet they are considered to be important by users.

49. See Vander Meer (1968) and Pasternak (1972) for a description of two such systems in Taiwan.

50. Similar procedures are reported in the traditional systems of south Bihar (Sengupta, 1980) and of Maharashtra (Kulkarni and Lele, 1980).

51. For instance, the Nan Hung project area of Taiwan, prior to modernization when all parts of the command area were unable to get puddling and transplanting water at the same time, planting used to be spread over as many as eight weeks and transplanting over four weeks. Harvesting however took place at about the same time (June) everywhere. (Vander Meer, 1968). This implies that part of the adjustment had to be made by choosing different varieties and /or accepting sizeable yield variation between the head reaches and the tail end of the system.

52. For concrete examples see Myers (1975); Pasternak (1972); Bacadayan (1973); Vaidyanathan and Janakarajan (1990).

53. See Nickum (1981: Reading 5) for a description of the complexity and sophistication of flexible scheduling in a multi-storage system.

54. In this connection mention may be made of an interesting custom in south Indian tanks (Chambers, 1977), Vaidyanathan and Janakarajan (1990); Vaidyanathan (ed.) (1998) where the responsibility for operating the sluice and distribution of water from village tanks is entrusted invariably to a member of the 'untouchable' caste: since he has no land and very low social status, he can never use his office to act arbitrarily. A similar custom is also said to have existed in parts of Japan.

55. Kelly (1980: 365–7) discussing the reasons why, despite its looseness, the Aka river system organization did not break down suggests that the following countervailing factors tended to limit conflict among water users and 'prevent disputes from deteriorating into serious destruction': (1) The irrigation relations were such that 'shifting lines of shared and conflicting interests created complex situational patterns of cooperation and conflict; (2) conflicts was probably mitigated by the disjunction of land tenure and irrigation: (i.e.) the distribution of cultivators and landowners over the water channel line; (3) 'Promotion of paddy land expansion through special concessions in surveying and registration resulted in wide disparities in tax burden, the undersurveyed, undertaxed lands on the tail end did not press their complaints of water shortage too forcefully'; (4) 'there was a fit between irrigation conflicts and administrative units'. Conflicts increased when the dissonance between landholding, cultivation, and residence increased.

56. For some idea of changes in particular villages, see Mizushima and Nara (1982), Guhan and Mencher (1982), Vaidyanathan and Janakarajan (1990).

57. The following description is largely based on Kathpalia (1980). See also Govt. of Andhra Pradesh (1982). More recent work on this aspect can be seen in Vaidyanathan and Janakarajan (1990), Rajagopal (1991), Ramanathan (unpublished), Meinzen Dick and Palanisami (1995), Wade (1981, 1982).

58 ... This is generally known as 'localization' and intended to ensure 'equitable' distribution of the benefits of water intensive crops (subject to soil and topography) within the command area, especially between the upper reaches and tail enders. For a discussion of the system and its working see, for example, GOI, PEO (1965); Govt. of Andhra Pradesh (1982: 53–65); and Govt. of Maharashtra (1962).

59. Numerous examples are available. See Pant (1981); GOI, PEO (1965); Govt. of Andhra Pradesh (1982).

60. The divergence between planned and actual crop pattern, and between planned 'localizations' and the actual distribution of water intensive crops within the command area, is a widespread phenomenon. For example, see GOI, Planning Commission (1965); Wade (1978); Hashim Ali (1980); Govt. of Andhra Pradesh (1982).

61. The difficulty in enforcing crop patterns and localization envisaged in the project is illustrated by the experience of the Lower Bhavani project in Tamil Nadu. See GOI, COPP (1965: 207).

62. There is reason to believe that this impression may be erroneous. There are reported instances where users who do not get enough water combine and try to persuade the canal officials, at considerable expense, to deliver them more water; or to distribute the available supplies among themselves equitably. Where they get no water at all, their option is often only to agitate for withdrawal of the water cess and betterment levy demanded of them (Wade, 1980 and Hashim Ali 1980). Wade (1978, 1980) reports one instance from Andhra Pradesh were rotational irrigation between different parts of the command area was successfully introduced at the initiative of the system manager in a drought year enabling much wider distribution of supplies and apparently without any significant reduction in yield. In another, the cultivation of paddy in unauthorized blocks was checked. These cases are interesting for two reasons: it shows that the experiment worked despite strong protests and resistance of upper reachers, because the system managers not only showed extraordinary initiative but also were willing and able to muster the power of the state to overcome the resistance. On the tail end question see also Hart (1978) and Pant (1981). Ramanathan, in his Ph.D. thesis (under preparation) has documented the success of tail enders in the Tambrapani system in securing significant increase in water allocations through a combination of political pressure and persuasion of bureaucracy.

63. In this context the distinction between unequal distribution between fields at different locations and between different users is important. In a gravity irrigation system it is the physical location of the field that has an important bearing on the ease of access to available supply. It appears unlikely that, in systems covering several thousand hectares, large farmers throughout, or even in a major part of, the command area will be concentrated in more favourable locations along the canal network. The probability of this happening in the normal course is very low. It appears to be difficult to manoeuvre the design of the system even at the distributatry level to systematically favour large owners. It is of course possible that after the system has been constructed, large farmers shuffle their holdings through purchase and sale with the explicit purpose of securing favourable location. This however needs to be established.

The importance of the size of landholding however is likely to be greater, and may even be decisive, in deciding how water is shared among farmers served by a particular outlet. It is noteworthy that, unlike the case of canals, where field location is more important, in the case of wells, the size of holding has a much more direct bearing on access to groundwater.

64. See for instance the papers presented at the All India Workshop on *warabandi* held at the Administrative Staff College, Hyderabad in 1980. Also Singh (ed.) (1982).

2

IRRIGATION IN INDIA: AN OVERVIEW OF DEVELOPMENT, 1950–90*

The expansion and improvement of irrigation facilities occupies a central place in India's programmes for agricultural development: between 1950 and 1997, the central and State governments together have directly invested nearly Rs 540 bn. on irrigation and flood control works which is by far the largest single item (accounting for a little under four-fifths) of the total public sector plan outlays on development of agriculture and allied activities. Substantial investments have been made by the private sector in developing groundwater irrigation and to make rain-fed lands fit for irrigated agriculture. The total extent of these investments has not been estimated. However, public sector financial institutions have provided farmers over Rs 70 bn. by way of loans at concessional rates for private irrigation investments.

In nominal terms, the public sector outlay has risen from an average of Rs 900 m. per annum during the First Plan to over Rs 650 bn. in the eighth. Part of this no doubt reflects the steep rise in prices and construction costs. But even after discounting for the price rise, plan outlays on irrigation during the seventh plan in real terms is estimated to be around three times that in the first (GOI, CWC, 1991).

The growth in the volume of investment has been accompanied by significant changes in the content of the programme. The marked shift of emphasis in favour of minor irrigation works during the Third Plan is reflected in a near doubling of their share in total irrigation outlays. Within minor irrigation, groundwater development by the farmers has received increasingly greater emphasis. The proportion of the government's direct

*This chapter draws heavily on my unpublished paper entitled 'Second India Series Revisited: Water' written at the MIDS at the request of the World Resource Institute, Washington DC.

investments in minor irrigation devoted to groundwater has probably increased. In any event, public sector financial support for private investment in minor irrigation (mostly for groundwater) has risen much faster than direct investment in minor irrigation. Another significant shift is the increased allocation to programmes for speedier and more efficient use of water. This is reflected in the emphasis on conjunctive use of surface and groundwater, command area development, and modernization of older irrigation systems (Table 1).

TABLE 1

Magnitude and Composition of Investment Through Plan Periods in the Irrigation and Flood Control Sectors in India, 1951–97

(Rs in crores)

| Plans | Major & medium irriga- tion | Minor irrigation | | | Command area develop- ment | Flood con- trol | Total (2+5+6+7) | Total public sector outlay (2+3+6+7) |
		Plan outlay	Institu- tional credit	Total (3+4)				
1	2	3	4	5	6	7	8	9
I Plan 1951–56	380	66	Neg.	66	NA	14	460	460
II Plan 1956–61	380	142	19	161	NA	49	590	571
III Plan 1961–66	581	328	115	443	NA	86	1110	995
Annual 1966–69	434	326	235	561	NA	44	1039	804
IV Plan 1969–74	1237	513	661	1174	NA	172	2583	1922
V Plan 1974–78	2442	631	780	1411	122	299	4274	3494
Annual 1978–80	2056	497	490	987	88	228	3359	2869
VI Plan 1980–85	7516	1802	1438	3240	521	596	11873	10435
VII Plan 1985–90	11343	3228	3312	6540	1428	942	20253	16941
Annual 1990–92	5320	1809	NA	1809	640	493	8262	8262
VIII Plan 1992–97	22415	5977	NA	5977	2510	1623	32525	32525
Total 1951–97	54104	15319	7050	22369	5309	4546	86328	79278

(Contd.)

(Table 1 Contd.)

| Plans | Major & medium irriga- tion | Minor irrigation | | | Command area develop- ment | Flood con- trol | Total (2+5+6+7) | Total public sector outlay (2+3+6+7) |
		Plan outlay	Institu- tional credit	Total (3+4)				
1	2	3	4	5	6	7	8	9
Ratio of Total/Total Public Sector outlay	68.2	19.3			6.7	5.7		100.0

Note: NA = Not available. Total (1951–97) excludes the institutional investments under Annual Plan (1990–92) and the Eighth Plan (1992–97); and also command area development outlays upto the IV plan (1969–74).

Source: GOI, Ministry of Water Resources 1989 a:108, 112–3

b: 2, A11, A12

GOI, 1992: 72, 86–91.

Given the important role of irrigation in the strategy of agricultural development and the large amount of resources devoted to it, this chapter attempts a critical review of the experience in terms of the spread of irrigation, its effectiveness and impact on agricultural productivity, the emerging problems and their implications for future strategy. The first section reviews the principal features of irrigation development since Independence in terms of the growth of irrigated area and quantum of water utilized from different sources, and its impact on agricultural productivity and the distribution of benefits from this. Some of the major problems that have surfaced during this period, the way in which they have been tackled, and their effect are then discussed. The concluding section deals with key issues concerning future development.

DEVELOPMENT AND IMPACT OF IRRIGATION SINCE 1950

Growth of Irrigated Area

An obvious and widely used index of irrigation development is 'area irrigated'. That there has been a significant increase in this is beyond doubt. According to the statistics of utilization of irrigation facilities compiled by the Planning Commission, the gross area irrigated has more than trebled over the past four decades: from 22.6 m. ha. in 1950–1 to 75.7 m. ha. in 1993–4. Over the same period the utilization of potential under major and medium irrigation works is estimated to have risen from

9.7 m. ha. to 29 m. ha.; the area irrigated by minor irrigation rose much faster from 12.9 m. ha. to 55 m. ha., much of it from groundwater. The area under minor surface works rose but marginally (Table 2).

Land use statistics (LUS), which are supposed to be compiled by the village revenue officials on the basis of first hand information, verified by field inspection of the actual position in every season, also show rapid growth of the total irrigated area from 22.6 m. ha. in 1950–51 to 68 m. ha. in 1993–4, but much less than estimated by the Planning Commission. The growing importance of groundwater in relation to surface water, and the stagnation of minor surface irrigation is also corroborated by the data on the extent of land served by different irrigation sources (Table 3).

There are however wide differences between the two estimates. According to the Planning Commission, between 1950–1 and 1993–4, gross irrigated area in the country as a whole increased by around 53 m. ha. while land use statistics show an increase of only 45 m. ha. (comparison by source is not possible because the Plan gives these figures only for the gross area irrigated and the LUS only for net area irrigated). There is however no consistent pattern in this difference across States. In some States the LUS figures are fairly close to the Plan estimates; in some (especially Haryana, Punjab and Rajasthan), the Plan estimates are well below the LUS data; but in others (notably Bihar, Uttar Pradesh and West Bengal), the Plan estimates substantially exceed that of land use and cropping. In fact, the bulk of the 'shortfall' in gross irrigated area compared to Plan estimates is on account of these three states (Table 4).

Both sources suffer from several weaknesses. Neither the Plan nor the LUS recognize and allow for conjunctive use of surface with groundwater. Since the Plan estimates utilization separately for surface and groundwater, there is the possibility of overestimating the total irrigated area due to double counting. Being based on various norms and assumptions rather than on verified ground reality, their estimates are also somewhat notional.

The accuracy of LUS data and their comparability across States and over time are also open to doubt. In the first place there is no uniformity in the agencies or procedures for recording irrigated area across States. In most of the southern States the data on all aspects of land use (including irrigation and crops) have been until recently compiled by local revenue officials, namely *patwaris/karnams*. In others, and this is the general practice in the north, the Irrigation Department is responsible for collecting the water cess and maintaining the record of area irrigated.[1] In yet others (Kerala, Orissa, West Bengal) the data are supposed to be collected through sample surveys. However, published data of irrigated area tend to repeat the same figure for several years, reportedly because they are very much

TABLE 2
Planwise Development of Irrigation (Potential/Utilization) in India 1950/51–1993/94

(Million ha.)

| | Major & medium irrigation | | Minor irrigation | | | | | | Total irrigation | | Gross Irri. area as per land utilization statistics |
| | | | Groundwater | | Surface water | | Total | | | | |
	P	U	P	U	P	U	P	U	P	U	
Pre Plan	9.7	9.7	6.5	6.5	6.4	6.4	12.9	12.9	22.6	22.6	22.6
1955-6	12.2	11.0	7.6	7.6	6.4	6.4	14.0	14.0	26.2	25.1	25.6
1960-1	14.3	13.1	8.3	8.3	6.5	6.5	14.8	14.8	29.1	27.8	28.0
1965-6	16.6	15.2	10.5	10.5	6.5	6.5	17.0	17.0	33.6	32.2	30.9
1973-4	20.7	18.7	16.4	16.4	7.0	7.0	23.5	23.5	44.2	42.2	40.3
1984-5	30.5	25.3	27.8	26.2	9.7	9.0	37.5	35.2	65.2	58.8	54.5
1993-4	33.8	29.3	41.5	38.6	11.8	10.5	53.3	49.1	84.8	76.4	68.1

Note: P = potential, U = utilization.
Source: GOI, Central Water Commission, *Water and Related Statistics 1998*, New Delhi.

TABLE 3

Growth of Net Irrigated Area Under Different Sources, All India from 1950/51–1993/94

(in million hectares)

	Canals				Groundwater				Total net irriga-ted	Gross irriga-ted
	Govt.	Private	Total	Tanks	Tube-wells	Other wells	Total	Other sources	area	area
1950–51	7.20	1.10	8.30	3.60	–	6.0	6.0	3.0	20.90	22.60
1955–56	8.0	1.40	9.60	4.40	–	6.8	6.8	2.2	22.80	25.60
1960–61	9.2	1.20	10.40	4.60	0.20	7.2	7.4	2.4	24.80	27.90
1965–66	9.8	1.10	10.90	4.40	–	8.6	8.6	2.5	26.40	30.90
1970–71	12.0	0.90	12.90	6.10	4.50	7.4	11.9	2.3	31.20	38.20
1975–76	12.9	0.90	13.80	4.0	6.80	7.6	14.4	2.4	34.60	43.40
1980–81	14.5	0.80	15.30	3.20	9.50	8.2	17.7	2.6	38.80	49.80
1990–91	16.1	0.30	16.10	3.30	14.20	9.90	24.10	2.80	46.30	60.70
1993–94	16.6	0.50	17.10	3.20	15.00	10.70	26.50	3.2	50.10	67.9

Source: Ministry of Agriculture, *Indian Agricultural Statistics*, GOI.

at variance with the potential and utilization estimated by the Irrigation Department.

There is also evidence pointing to substantial errors in the data compiled by village officials. A sample survey conducted in the fifties by the Programme Evaluation Organization of the Planning Commission showed that the total irrigated area as a proportion of cultivated (and cropped) area as reported by the sample households differed substantially from that shown in village records. For the country as a whole, the latter was lower: the village records showed 23.8 per cent of the net cultivated area as irrigated while the information provided by sample households gave a figure of 27.2 per cent. The direction and extent of difference however varied widely among States (Government of India, 1961). More recently, in Andhra Pradesh, considerable discrepancies were found in the estimates of irrigated area for selected project command areas compiled by different agencies even though all of them relied on the same primary source, namely, the village revenue officials (Government of Andhra Pradesh, 1982, Ch. 3). The system of joint inspection by the Revenue and Irrigation officials, which is meant to reconcile differences in estimates, has not in fact worked satisfactorily.

TABLE 4

Statewise Area Irrigated by Surface and Groundwater as Per Land Use Statistics and the Planning Commission

(Area '000 ha.)

State	Land Use Statistics[1]												Planning Commission[2]								
	1960-1 Net irrigated area by				1973-4 Net irrigated area by				1992-3 Net irrigated area by				1960-1 Gross irrigated area by			1973-4 Gross irrigated area by			1991-2 Gross irrigated area by		
	Surface water	Groundwater	Total	GIA	Surface water	Groundwater	Total	GIA	Surface water	Groundwater	Total	GIA	Surface water	Groundwater	Total	Surface water	Groundwater	Total	Surface water	Groundwater	Total
Andhra Pradesh	2561	347	2908	3474	2636	645	3281	4223	2618	1411	4029	5085	2558	425	2983	3297	775	4072	4353	1528	5881
Bihar	1703	358	2061	2063	1604	716	2320	2797	1643	1701	3344	4040	1524	260	1784	2111	800	2911	3968	3105	7073
Gujarat	116	567	683	734	282	1119	1401	1591	586	2056	2642	3227	162	620	782	512	1200	1712	1148	1648	2796
Haryana*					1018	718	1736	2583	1389	1239	2628	4472	442	300	742	1511	1010	2521	1912	1443	3355
Karnataka	725	132	857	977	928	273	1201	1421	1469	725	2194	2802	951	150	1101	1178	325	1503	2089	706	2795
Kerala	316	2	318	513	351	5	356	637	269	66	335	376	528	neg	528	644	5	649	1049	87	1136
Madhya Pradesh	599	324	923	939	969	675	1644	1732	2453	2322	4775	4918	941	330	1271	1221	700	1921	2569	1456	4025
Maharashtra	477	595	1072	1220	632	839	1471	1763	1122	1348	2470	3235	665	680	1345	887	925	1812	2002	1508	3510
Orissa	938	38	976	1216	743	135	878	1260	1236	834	2070	2471	1022	neg	1022	1438	90	1528	2290	536	2826
Punjab	2199	895	3094	3887	1296	1680	2976	4618	1463	2398	3861	7142	2186	900	3086	2466	2600	5066	2632	3174	5806
Rajasthan	737	1014	1751	1752	1115	1262	2377	2678	1667	2804	4471	5486	765	1020	1785	1394	1400	2794	2094	1914	4008
Tamil Nadu	1883	579	2462	3237	1889	926	2815	3674	1497	1201	2698	3385	1872	785	2657	1971	1000	2971	2156	1242	3398
Uttar Pradesh	4807	267	5074	5531	3034	4207	7241	8491	3684	7638	11322	15996	3606	2800	6406	4514	5300	9814	7109	16339	23448
West Bengal	1181	187	1368	1368	1169	320	1489	1541	1199	712	1911	1911	1668	neg	1668	9428	250	9678	2765	1159	3924
All India	17506	7154	24660	28020	17969	13217	31186	39009	22295	26455	48750	64546	28005	8270	36275	25890	16391	42281	39576	36054	75630

*Included in Punjab.

Note: As per land use statistics surface irrigation is sum of canal, tank and other sources, Groundwater includes both wells and tubewells. As per Planning Commission surface water includes Major and Medium and Minor surface works.

Source: (1) GOI, Indian Agricultural Statistics Ministry of Agriculture; (2) GOI, Central Water Commission, 1992.

Where area classified as irrigated carries a higher assessment (on account of higher revenue assessment or liability to pay water charges), the irrigator has an incentive to under-state the irrigated acreage. In principle, the village officials should be interested in recording the correct acreages in the interests of revenue. They are however not free from the pressures from farmers among whom they themselves often figure prominently. This bias is however more likely to affect the area reported under specific crops than the total irrigated area. There is a high probability that area under crops carrying a relatively high rate of water charges or other cesses are underestimated. On the other hand, where, as in the case of lands served by wells and other private works, the fact of irrigation has no implications for government revenue, the *patwaris* may not take as much care in recording the area and crops irrigated.

The quality of these data depends crucially on the care with which the village level records on land use and cropping are scrutinized and verified by inspecting officials. Whatever may have been the situation in the past, there is little doubt that the sharp decline in the importance of land revenue and the enormous increase in the tasks entrusted to revenue officials has led to greater laxity in the compilation of these data and even more in their verification. The situation has further deteriorated in States that have replaced the traditional system of appointing residents as *patwaris/ karnams* by a cadre of transferable functionaries. Hence the available data leave much to be desired both conceptually and in terms of reliability.

Conceptually, apart from adopting the many sensible suggestions of the AP Commission, for clear and unambiguous definition of 'utilization', it is desirable to introduce a more elaborate classification of sources to distinguish between single and multiple sources of irrigation and between different combinations of sources. Also, the crop area and cropping intensity data would be much more useful in assessing the impact of irrigation if they were available separately for land that forms part of the command area of one or other irrigation source and those that are not. Equally important is a mechanism for generating reliable data on a regular basis. Since the traditional arrangements for this purpose are in considerable disarray, alternatives need to be developed.

Sample surveys conducted regularly by independent professional agencies to verify the reports of achievements in different categories of works received from implementing agencies are essential.[2] The scope of such surveys need to be enlarged to cover the extent of water delivered, areas and crops actually irrigated, the techniques of cultivation and yield obtained in surface projects and, in the case of ground water, the number of wells/tube wells,·volume of water extracted, costs, and productivity. The Planning Commission could make greater use of the National Sample Survey, and the data from the Cost of Cultivation Surveys, for this purpose at a macro level. These however need to be supplemented by surveys of selected projects of different types to obtain more detailed region-specific information. These surveys are likely to be more effective and economical if they are entrusted (with funding) to independent research institutions all over the country on a continuing basis. The state governments and the financing bodies (e.g. NABARD) could also participate in such surveys.

Remote sensing through satellite imagery has greatly enhanced the capability to gather information, at frequent intervals, on forest cover, flooding, level of storage in reservoirs, moisture stress on cultivated land, the extent and location of various types of degraded land. This capability is only beginning to be utilized to monitor changes in forest cover. But applications to other aspects that are relevant both for short-term management (as in the case of canal management, assessment of the true magnitude and severity of flood or drought), and for long-term planning of resource use are as yet limited. The technical capability exists in the National Remote Sensing Agency and several universities. However, the executive agencies of government, which need the data and have the funds, do not take active and sustained interest in utilizing this capability and strengthening it. Archaic and pointless restrictions on access to and use of remote sensing data are a serious impediment that needs to be overcome.

EFFECTIVENESS OF WATER USE

Utilization of Created Potential

The expansion of irrigated area per se is a poor measure of irrigation development inasmuch as it tells us little about the speed and effectiveness with which irrigation facilities are utilized. The persistent gap between the estimated potential for irrigation (especially under major and medium projects) and area reported to be actually irrigated has attracted much attention. The Planning Commission estimates (see Table 2) show that between 1950 and 1993 utilization of irrigation from all sources rose by 48–9 m. ha. while potential rose by 56–7 m. ha. This implies that about one-eighth of the additional potential has not been used.

The gap is much larger (33 per cent) when we compare LUS estimate of gross irrigated area with Planning Commission estimates of potential. The additional utilization as estimated by the Commission was 83 per cent of additions to potential in the first two Plans; the ratio improved to

94 per cent during 1961–74 but fell again to 85 per cent during 1978–90. In 1960–1, LUS estimates of gross irrigated area was 96 per cent of total potential estimated by the Planning Commission; the ratio has been falling progressively since to reach 91 per cent in 1973–4 and 78 per cent in 1989–90.

These comparisons are however neither particularly reliable or meaningful. The deficiencies in both Planning Commission estimates of utilization and the LUS estimates of gross irrigated area have already been discussed at length. Estimates of 'potential created' are equally faulty. Part of the problem lies in ambiguities in the definitions of 'potential' as well as differing interpretations of the guidelines. Estimates of potential gross irrigated area rest on assumptions regarding the likely water availability, the water requirements of crops, crop patterns and the extent of losses in conveyance and application of water. The validity of these assumptions is open to doubt.

A detailed study in Andhra Pradesh found none of the engineers examined by them 'were able to give the actual basis for the duties adopted' (in the design). Nor did they find evidence that differences in climate, topography, soils, and other conditions from one project to another had been taken into account in determining water requirements. The parameters regarding losses and system efficiency generally were not based on systematic investigation and the actual losses often proved to be much larger than assumed in the project. There was no correspondence between the crop pattern and dates of sowing assumed at the stage of initial planning and the actual crop pattern and cultivation schedules adopted by the farmers.

The projects therefore deliver water at the offtakes and outlets with little relevance to the actual water requirements of the crops grown. This contributes to a gap in the planned irrigation potential as well as gap in utilization of the potential. [GOAP 1982: 17.]

Analysis of eight projects by this Commission, which is unique in the country, showed that, though the volume of water actually supplied exceeded allocations in several cases, the actual irrigated area was 11 to 70 per cent short of potential. A part of the gap (10 to 46 per cent) arose from the divergence of actual from planned crop patterns, but even after correcting for it, utilization was substantially below the potential (GOAP, 1982: 46–7).

Conditions of course vary from State to State. While the Andhra Pradesh experience may not hold in every respect or in the same measure everywhere, the problems highlighted by the Commission appear to be quite widespread. Numerous examples can be cited from all over the country corroborating the above observations, especially regarding the tendency for actual crop patterns to be considerably more water intensive than planned.

In the case of minor works, there is some indication that allowance is made for depreciation in estimating 'additions to potential' which is described (at least since the early 1970s) as new irrigation *minus* depreciation of existing works. Benefits from 'schemes for stabilization of irrigation over areas where irrigation was previously available but is now made more certain through supplemental sources' are treated separately and not counted in addition to potential. There is however little information about the basis on which depreciation estimates and the discrimination between new irrigation and stabilization of existing sources are made; or on how this information is monitored on the ground.

A particularly important issue in this context is the treatment of groundwater used in conjunction with surface water. If allowance is not made for this overlap, there will be some double counting in estimates of irrigation potential. The success of the government's policy of actively encouraging conjoint use will, if anything, have accentuated this bias over time. Nor does the Planning Commission take into account the effect of falling water tables and growing number of wells/tubewells and pumpsets on water extraction or area irrigated per well/tubewell. There is no attempt to revise these parameters on the basis of systematic surveys from time to time. Clearly if, as seems likely, water extraction per well is falling, total area irrigated by groundwater, as estimated from the number of wells, pumpsets and past norms of area per well will have a further upward bias. The Planning Commission has taken a long time to recognize these problems. It is only in the Eighth Five Year Plan document that we find an explicit statement on them.

There are differences in the interpretation of the concept by the reporting agencies. The system of monitoring and verifying information provided by executive agencies on both potential and actual irrigation is inadequate and casual and there are substantial reporting and compilation errors in the data. The traditional system of compiling data on land use, irrigation and cropping by the village *patwari* which in principle provides an independent source of data on actual has deteriorated.

'The potential area which can be irrigated in a system depends on several factors including, besides the availability of distribution networks, the volume and seasonal pattern of water supply, the losses in conveyance, distribution and application, the extent to which the conjunctive use is developed and the crop pattern on ground. In so far as the the assumptions in respect of these parameters underlying the project design are not realized in full, there is bound to be a divergence between actual area irrigated and potential created. There is considerable evidence to show that the crop patterns actually adopted by farmers are often much more water intensive than assumed and this is one important reason why actual area irrigated is smaller than designed potential. In judging the performance of irrigation systems, we should move away from potential or actual area irrigated and focus instead on the amount of water made available through a system (both overall and in each main crop season), the frequency and predictability of water supply and its effect on cropping intensities, crop patterns and yields. [GOI, PC, 1992.]

This understanding is, however, yet to be reflected in the way irrigation performance is in fact monitored.

Trends in Water Utilization

Data on utilization of surface water are much harder to come by: in principle, since all major and medium surface works are managed by government personnel and are equipped with guages to measure flows at the offtake points of dams, diversion wiers, as well as at the head of main and branch canals, fairly accurate measurements of total water drawn should be relatively easy to compile at the level of individual projects and aggregated at the regional, State or basin level. These data are supposed to be recorded practically on a daily basis, but are apparently not compiled and collated for all projects. Some States (e.g. Tamil Nadu) have been publishing data on the quantum of water releases from select reservoirs. Figures of total releases from all reservoirs are not published.

A rough idea of the level and pattern of water utilization in the country as a whole and for major river basins can however be gleaned from figures given in the Report of the Irrigation Commission (GOI, 1972) and in some recent publications of the Central Water Commission (Reddy, 1990) (Table 5). The gross withdrawal of water from the surface and from underground is estimated to have increased from 376 bn. m^3 in 1968–9 to 549 bn. m^3 in 1990. Withdrawal for non-agricultural uses has increased much faster than average; their share in gross utilization is however only 12 per cent. Since much of it is available for reuse, net non-agricultural consumption accounts for a miniscule portion of the total. Most of the water is used for agriculture and livestock.

Gross utilization for irrigation and livestock, according to available estimates of the Central Water Commission, rose from an estimated 360 bn. cu. m in the late sixties to 490 bn. cu. m in 1990—an increase of about one third. The reported increase (100 per cent according to the Planning

Commission and 60 per cent according to land use data) in gross irrigated area during this period is considerably greater than the estimated increase in volume of water use for irrigation. The implied decrease in the quantum of water use per unit irrigated area may be partly explained by the shift in irrigated cropping patterns away from water intensive crops (paddy and sugarcane) to other crops whose water requirements are smaller (Table 6). An increasing proportion of the supply is drawn from groundwater, both as a sole source and by recycling seepage in surface irrigated areas. Conveyance and application losses in groundwater irrigation is much lower

TABLE 5

Estimated Water Utilization by Use

(Million ha. metres)

	1968–69*				1990**	
	Gross	%	Net	%	Gross	%
Domestic	0.91	2.9	0.47	2.1	2.50	3.5
Industrial	0.27	0.7	0.02	0.1	1.50	2.1
Thermal Power	0.69	1.8	neg	neg	1.90	2.6
Agriculture and Live Stock	35.68	95.0	21.58	97.6	49.30	91.8
Total	37.55	100.0	22.08	100.0	55.20	100.0

Source: *Chaturvedi, 1976.
** Reddy, 1992.

TABLE 6

Irrigated Crop Pattern in the Early 1960s, 1970s, and 1980s
(percentage of gross irrigated area)

	1960–1 to 1962–3	1970–1 to 1972–3	1987–8 to 1989–90
Rice	45.30	37.00	27.70
Wheat	15.30	26.90	31.70
Total Cereals and Millets	72.00	73.60	73.00
Total Pulses	7.00	5.00	3.60
Sugar-Cane	5.60	4.70	5.20
Fruits and Vegetables	1.70	2.80	4.10
Cotton	3.60	4.10	4.10
Total Oilseeds	1.60	2.90	7.30
Other Crops	8.30	6.70	2.70
Total	100.00	100.00	100.00

Source: GOI, Ministry of Agriculture, *Indian Agriculture Statistics* (various issues).

and recycling in effect increases the proportion of water supplied from surface sources that is effectively used. Since both factors increase the ratio of net use to gross utilization, net effective use of irrigation water may have risen faster than gross utilization.

The confidence with which this important conclusion can be asserted is however considerably weakened because the basis of the utilization estimates (in terms of data sources, assumptions etc.) are not spelt out anywhere. Nor do we have disaggregated estimates along comparable lines for river basins and/or States. Chaturvedi (1976) reports basinwise estimates of surface and groundwater withdrawals in 1968–9 given in the report of the Irrigation Commission. The latest published estimates relating to 1989 (which are incomplete) suggest that utilization excluding the Ganga and Brahmaputra has risen by 20 per cent (Table 7).[3]

Estimates of groundwater use, based as they are on norms and assumptions rather than direct measurements of actual extraction, are even more suspect. Nor do we know how much of the groundwater come from conjoint use. In the absence of this information it is impossible even to judge whether the data on level and source of water consumption use and area irrigated are broadly consistent with each other, not to speak of assessing the change in the effective water use per unit area, the factors underlying it and the significance for efficiency of water use.

PRODUCTIVITY IMPACT OF IRRIGATION

The characteristics of water supply to irrigated areas are important because they determine the impact on productivity of land. It is well known that irrigation makes a difference to land use patterns (especially the extent of fallows), the intensity with which land is cultivated, crop patterns, the level of fertilizer use and the efficiency with which nutrients are used. The extent of irrigation is however not the only factor; the quality of irrigation (in terms of quantum, assurance, and timeliness of water supply) as well as agro-climatic conditions have a significant bearing on the difference that irrigation makes to productivity.

Irrigation quality is partly a function of the nature of the water source, partly the way it is managed. Broadly speaking, tanks and other water surface works that depend mostly upon local rainfall offer inherently limited scope for augmenting soil moisture, in terms of their reliability and flexibility in adjusting supplies to the needs of crops in specific locations and at specific times. In all these respects large storages that tap water from a wide catchment are superior; but groundwater is unquestionably the best source. How far the potential quality of each source

TABLE 7
Estimates of Water Potential and Utilization 1976 and 1992
(million hectare meters)

River basins	Annual precip- tation	Annual surface runoff		Annual groundwater recharge		Ultimate Potential				Actual Utilization				Capacity of 3 large reservoirs	
						Surface		Groundwater		Surface		Groundwater			
		1	2	1	2	1	2	1	2	1	2	1	2	1	2
Ganga	116	55.01	52.50	14.62	25.00	18.50	25.00	10.60	17.2	13.20	N.A	4.10	4.86	3.35	3.70
Brahmaputra	122	42.20	53.70	3.97		0.88	2.40	1.80	2.08	0.61	N.A	0.02	0.08		
Barak & other Rivers	286	17.50	6.00	1.60		0.37		0.66	0.13	0.10	0.10	0.07	0.00	0.01	0.11
West flowing rivers below Tapi	279	21.79	19.90	2.01		2.88	3.62	1.04	0.90	1.27	1.14	0.08	0.33	1.40	1.73
Tapi including	78	1.97	1.80	0.61		1.46	1.45	0.41	0.59	0.54	0.67	0.16	0.20	0.81	0.85
Narmada	121	4.01	4.13	1.24		2.96	3.45	0.71	0.92	0.28	1.00	0.14	0.13	0.26	0.66
Mahi including Dhadhar	83	1.18	1.18	0.35		0.87	0.31	0.25	0.44	0.39	N.A	0.12	0.20	0.43	0.47
Sabarmati	76	0.38	0.41	0.27		0.27	0.19	0.06	0.44	0.15	0.18	0.01	0.13	0.10	0.13
Luni and others Saurashtra & Kutch	38	1.23	1.51	0.86		0.91	1.50	2.01	1.26	0.38	0.25	0.84	0.48	0.18	0.43
Indus	56	7.69	7.33	2.90		4.93	4.60	1.12	2.34	4.60	4.00	0.85	1.82	1.44	1.45
East flowing rivers between Ganga and Mahanadi	147	4.35	4.70	1.63		4.09	2.51	0.65	0.62	0.02	N.A	0.05		0.43	0.54

(*Contd.*)

(Table 7 Contd.)

River basins	Annual precipitation	Annual surface runoff		Annual groundwater recharge		Ultimate Potential				Actual Utilization				Capacity of 3 large reservoirs[3]	
						Surface		Groundwater		Surface		Groundwater			
		1	2	1	2	1	2	1	2	1	2	1	2	1	2
Mahanadi	146	7.07	6.69	2.13		6.64	5.00	1.06	1.40	2.13	1.80	0.03	0.10	0.79	0.85
East flowing between Mahanadi and Godavari	111	1.72	1.77	0.73		1.61	1.31	0.40	1.17	0.88	N.A	0.02	0.07	0.08	0.12
Godavari	110	11.54	11.90	3.31		8.53	7.63	1.92	3.10	3.27	3.80	0.54	0.60	1.49	1.95
Krishna	81	5.78	6.78	2.65		5.78	5.80	1.76	2.08	5.24	4.70	0.65	0.63	2.99	3.45
Pennar and other east flowing rivers between Pennar and Kanyakumari	82	2.53	2.46	1.82		2.70	2.36	4.00	2.07	2.19	0.50	0.46	1.05	0.25	0.18
Cauvery	99	1.86	2.10	1.23		2.06	1.90	1.00	1.04	1.79	1.80	0.26	0.58	0.55	0.74
East flowing below Ganga	91	0.90		0.54		1.17				0.96		0.12			
Total	118	188.75		42.41		66.60	69.33	26.10	37.8	38.60		7.90	11.52	14.70	17.37

Note: (1) Relates to 1969–74, Chaturvedi, 1976. (2) Latest published figures as reported in GOI, Central Water Commission, *Water Related Statistics 1998*. Relates to late 1980s and early 1990s. (3) Reservior capacity figures for 1969–70 include those under construction figures for period 2 relate to completed projects.

is realized depends crucially on the sophistication of physical design and the quality of management.

The impact is also conditioned by the level and distribution of rainfall in relation to that of evapo-transpiration which in turn is a function of agro-climatic conditions. Other elements of biochemical technology—namely the seed varieties, level of fertilizers and other inputs—also have a significant bearing on the outcome. Irrigation has a strong synergise effect both on the scope for using these inputs and the efficiency with which they are used by the crops. But the quality of both water and agronomic management are crucial.

Cropping Intensity

The beneficial impact of irrigation on overall cropping intensity has been underlined by several studies (Dhawan, 1992; Ray, 1992; Vaidyanathan et al., 1996 being recent examples). Dhawan found, using the national time series data for 1950–1 to 1986–7, that variations in cropping intensity are significantly affected by both the irrigation ratio and rainfall. The positive impact of rainfall and irrigation crop intensity is confirmed by cross-section analysis for different points of time. In addition, it appears that the impact of rainfall tends to decrease as rainfall rises, and that greater irrigation in the *kharif* (monsoon) season tends to dampen the effect of the irrigation ratio per se on cropping intensity. Analysis of district level data for Tamil Nadu (Vaidyanathan et al., 1996) also show irrigation and rainfall to be significant determinants of cropping intensity: the higher the rainfall and the higher the irrigation ratio, the higher the cropping intensity. However, neither the seasonal distribution of rainfall nor the proportion of net irrigated area served by wells were found to have any significant impact on cropping intensity.

There is room for further refinement of both data and analysis of the cropping intensity–irrigation relation. Quite apart from errors and biases in the data, the conventional measure of cropping intensity, namely the ratio of gross cropped area to net cultivated area, does not take into account differences in the duration of crops: some seasonal crops take only 90 days from sowing to maturity while, at the other extreme, crops like surgarcane take a year or even more to mature. A better measure would be in terms of the number of months during which crops occupy from a given extent of land. Some attempts have been made to measure cropping intensity in this refined manner and assess its determinants (Ray, 1992). It is also useful to study the variations in cropping intensity separately on irrigated and unirrigated lands, and land served by different sources of irrigation and their combinations. Suitable reclassification of existing data (from LUS and special surveys) can take us some way towards dealing with these issues, but fresh surveys to obtain more reliable and conceptually satisfactory data will also be needed.

Crop Pattern

In general paddy (and wheat in some areas), sugar-cane, fruits and vegetables figure more prominently in irrigated areas; while coarse cereals, pulses and oilseeds, and cotton are conspicuous in rain-fed cropping (Table 6). There are regional variations in crop patterns as between these two categories of areas. There have also been significant changes in both irrigated and rain-fed crop patterns all over the country (see Table 8).

Rice and wheat, which accounted for about 60 per cent of gross irrigated area in the early 1960s, constituted about 59 per cent in the late 1980s (Table 6). There has however been a major shift from rice to wheat in the irrigated crop patterns during this period. Rice being a water intensive crop, the reduction in its share of irrigated crop area would reduce the use of irrigation water per unit of total irrigated area. This may however be offset by an increase in the proportion of area under water intensive, annual crops like sugar-cane and banana (included under fruits and vegetables). Official data do not show an increase in the share of sugar-cane, but for reasons cited, the extent of increase may be understated. Data on other water intensive crops are not readily available.

Another way of looking at the crop pattern is in terms of the extent of its diversification measured by the sum of deviations of the actual shares of each crop from the shares that would obtain under a perfectly diversified pattern (i.e. a pattern in which the available area is equally distributed among all crops). Taking the country as a whole, the unirrigated crop pattern was somewhat more diversified in comparison to the irrigated crop patterns in the early 1970s. The irrigated crop pattern has become more diversified while that of the unirrigated areas has become more concentrated. In half of the States unirrigated areas have a more diversified pattern than irrigated areas in both periods. In a majority of States, the unirrigated crop patterns become less diversified. In most States irrigated cropping become more diversified (Table 8).

An ongoing study of Tamil Nadu (P. Anbazhagan, working on Ph. D. thesis at MIDS) further shows that the degree of crop diversification is strongly influenced by the source of irrigation and that, controlling for the source of irrigation, there is a noticeable tendency for irrigated crop patterns to become more diversified in most districts of the state even as rain-fed crop patterns are becoming less so. Given the deficiencies in the underlying data, and the fact that the grouping of crops into broad

TABLE 8
Percentage Distribution of Irrigated and Unirrigated Gross Crop Areas, 1970 to 1972 and 1987 to 1989, All India and Major States

State	Year	Source	Paddy	Wheat	Other cereals	Pulses	Oil-seeds	Sugar-cane	Condi-men & spices	Fruits & veg.	Cotton	Tobacco	ONFC	Index on con-centra-tion*
Andhra Pradesh	1970–72	UI	2.14	0.11	48.03	15.81	23.06	0.02	2.74	2.23	3.38	2.25	0.22	119.3
		I	77.56	0.24	7.01	0.04	5.91	3.18	2.88	1.62	0.56	0.81	0.20	137.1
	1987–89	UI	2.30	0.04	29.68	19.59	31.42	0.03	2.42	5.88	7.09	1.25	0.31	105.0
		I	75.36	0.16	3.85	0.35	9.46	3.91	3.56	0.39	1.46	0.91	0.56	133.3
Bihar	1970–72	UI	45.24	9.50	17.98	20.19	3.17	1.41	0.42	1.90	0.03	0.13	0.02	113.1
		I	62.68	27.35	2.81	0.69	0.43	1.41	0.23	4.21	0.01	0.11	0.06	143.7
	1987–89	UI	54.94	5.81	10.97	19.00	3.32	1.60	0.20	3.87	0.00	0.08	0.12	119.2
		I	45.21	41.49	6.63	0.37	0.53	0.62	0.14	4.66	0.00	0.25	0.07	137.1
Gujarat	1970–72	UI	4.24	1.79	42.21	5.51	25.29	0.00	0.01	0.37	19.82	0.52	0.25	120.1
		I	10.59	25.66	9.48	0.54	5.68	2.93	7.03	4.89	28.56	3.19	1.43	75.8
	1987–89	UI	4.86	1.47	39.80	12.97	29.76	0.00	0.01	0.21	10.28	0.66	0.00	112.9
		I	10.42	17.63	10.20	2.66	20.68	4.87	5.94	6.36	15.68	3.51	1.99	58.4
Haryana	1970–72	UI	1.43	8.78	46.04	36.95	5.46	0.90	0.04	0.20	0.15	0.00	0.05	129.6
		I	11.61	45.64	10.12	12.09	2.32	5.14	0.55	1.41	10.52	0.03	0.57	89.1
	1987–89	UI	0.42	3.59	54.75	31.08	8.76	0.62	0.00	0.24	0.33	0.02	0.21	135.3
		I	15.15	47.39	6.92	4.93	7.86	3.51	0.24	1.61	11.78	0.00	0.65	94.1
Karnataka	1970–72	UI	4.52	3.48	45.96	16.48	14.30	0.01	1.73	1.08	11.57	0.35	0.50	103.9
		I	51.80	2.25	19.38	0.31	6.10	7.18	3.86	3.86	3.60	0.21	1.44	106.1
	1987–89	UI	4.71	1.87	41.99	17.39	23.43	2.04	1.84	5.02	0.01	0.50	1.20	111.1
		I	28.95	3.13	20.23	2.86	19.32	3.27	4.02	5.79	9.42	0.09	2.89	83.2

(Contd.)

(Table 8 Contd.)

State	Year	Source	Paddy	Wheat	Other cereals	Pulses	Oil-seeds	Sugar-cane	Condi-men & spices	Fruits & veg.	Cotton	Tobacco	ONFC	Index on con-centra-tion*
Madhya Pradesh	1970–72	UI	20.86	15.47	25.45	22.95	10.54	0.02	0.29	0.38	3.70	0.01	0.34	99.6
		I	39.13	37.42	2.80	7.39	1.45	3.20	2.76	3.86	0.72	0.09	1.17	116.8
	1987–89	UI	21.75	10.95	23.42	23.24	17.46	0.01	0.32	0.44	2.41	0.00	0.00	102.7
		I	27.46	42.77	1.04	11.69	4.59	2.07	3.65	3.41	2.99	0.01	0.26	109.2
Maharashtra	1970–72	UI	6.49	3.58	47.38	14.52	10.81	0.00	0.70	0.22	16.22	0.07	0.02	105.1
		I	21.48	18.05	21.17	3.50	2.02	13.26	4.67	10.91	4.57	0.04	0.35	78.8
	1987–89	UI	6.28	2.10	47.28	16.17	11.83	0.01	0.43	0.43	15.42	0.05	0.00	108.7
		I	19.24	15.94	20.61	5.28	5.27	14.83	3.85	10.92	3.72	0.01	0.30	72.1
Orissa	1970–72	UI	64.37	0.26	7.20	14.76	5.26	0.27	1.07	6.35	0.01	0.25	0.00	122.3
		I	75.40	1.13	1.89	4.31	5.32	1.28	0.00	1.43	0.00	0.00	9.32	133.1
	1987–89	UI	39.85	0.03	7.66	27.75	15.05	0.00	1.26	8.12	0.00	0.19	0.00	110.8
		I	66.71	1.78	1.05	4.62	4.38	2.01	3.31	15.35	0.00	0.08	0.69	127.8
Punjab	1970–72	UI	3.17	27.09	22.61	24.59	18.34	1.37	0.09	1.05	0.96	0.02	0.73	112.5
		I	10.58	55.56	12.55	2.93	2.94	2.51	0.46	1.29	11.79	0.00	0.39	106.2
	1987–89	UI	4.63	33.25	25.60	19.38	8.36	1.44	0.03	0.54	0.83	0.00	5.89	102.1
		I	28.98	49.44	2.93	1.26	2.09	1.55	0.08	1.51	11.38	0.00	0.74	125.1
Rajasthan	1970–72	UI	0.74	3.72	57.78	26.52	9.27	0.02	0.72	0.05	0.81	0.38	0.00	132.6
		I	1.60	40.34	20.26	11.12	4.35	1.22	3.83	1.20	8.04	0.13	1.11	95.7
	1987–89	UI	0.68	1.51	62.22	22.40	11.73	0.01	1.20	0.06	0.20	0.01	0.00	138.1
		I	1.14	37.35	10.95	6.37	26.13	0.47	6.28	1.51	8.43	0.04	1.27	94.3
Tamil Nadu	1970–72	UI	5.54	0.00	40.18	14.67	28.79	0.00	1.47	3.79	5.41	0.00	0.16	111.7
		I	71.12	0.00	9.60	0.36	6.45	3.51	2.35	2.92	2.95	0.33	0.41	127.5
	1987–89	UI	5.17	0.00	34.01	19.55	27.44	0.00	2.88	5.32	4.56	0.00	1.06	107.5
		I	59.77	0.00	5.31	1.33	13.96	7.17	1.94	5.43	3.71	0.30	1.03	101.9

(Contd.)

(Table 8 Contd.)

State	Year	Source	Paddy	Wheat	Other cereals	Pulses	Oil-seeds	Sugar-cane	Condi-men & spices	Fruits & veg.	Cotton	Tobacco	ONFC	Index on con-centra-tion*
Uttar Pradesh	1970–72	UI	9.40	13.90	48.51	19.58	4.52	3.12	0.08	0.72	0.06	0.02	0.10	110.8
		I	9.34	50.75	12.85	11.60	0.96	10.85	0.20	2.69	0.54	0.11	0.12	99.8
	1987–89	UI	18.68	6.30	54.27	14.17	3.03	1.84	0.08	1.57	0.01	0.00	0.04	119.7
		I	11.88	42.11	27.54	3.83	2.13	8.23	0.31	3.33	0.09	0.07	0.17	109.1
All India	1970–72	UI	13.62	7.03	39.64	17.95	12.43	0.83	377.00	0.88	6.35	0.32	0.18	94.5
		I	33.96	30.36	12.21	5.39	3.24	5.54	1.61	3.01	4.91	0.32	0.46	96.5
	1987–89	UI	14.94	4.21	38.21	19.13	15.14	0.41	1.17	1.72	4.66	0.20	0.21	102.1
		I	27.69	31.73	13.57	3.63	7.26	5.23	1.73	4.09	4.13	0.30	0.64	91.4

Note: I stands for Irrigated and UI stands for Unirrigated crops.
ONFC stands for Other non-food crops.
*The index of crop concentration is the arithmetic sum of the deviations of the share of each crop in the total from the mean share.

Source: Agricultural Statistics of India, Ministry of Agriculture, Various Issues.

categories tend to understate the extent of diversification, especially on rain-fed land, these inferences must be treated with caution.

Overall Impact on Productivity: Indirect Estimates

The cumulative impact of irrigation on cropping intensity, crop patterns, and individual crop yields is reflected in the overall productivity of irrigated in relation to that of rain-fed areas. Distribution by crop of the total irrigated and unirrigated areas are published regularly. Separate estimates of irrigated and unirrigated yields (based on crop cutting surveys) are also available State-wise for some major crops. For others only overall average yield estimates are published. Many of them, especially those that are quantitatively significant, happen to be predominantly irrigated or predominantly rain-fed. It is therefore possible, even with the available data, to estimate the aggregate and per hectare output of irrigated and unirrigated crops. Some attempts have been made in this direction (Dhawan, 1988; Vaidyanathan, 1987; and Vaidyanathan et al., 1994).[4] The results of the latest exercise of this genre giving statewise estimates of the value of output (at constant prices of 1980–3) per hectare of irrigated and unirrigated crop area are set out in Table 9.

TABLE 9
Productivity of Irrigated and Unirrigated Areas by States

	Average output per ha. of gross cropped, 1979–83 (Rs/ha. at 1980–83 Prices)			Trend in output per ha. 1970–1983	
	Irri.	Unirri.	Ratio of I/UI	Irri.	Unirri.
Andhra Pradesh	6689	2282	2.93	+S	+S
Bihar	2993	2278	1.31	NS	+S
Gujarat	6353	2714	2.34	NS	NS
Haryana	4500	1293	3.48	+S	NS
Karnataka	6825	2297	2.97	+S	NS
Madhya Pradesh	3391	1656	2.05	+S	NS
Maharashtra	7415	1603	4.63	NS	–S
Orissa	3958	2188	1.81	+S	+S
Punjab	5998	2628	2.28	+S	NS
Rajasthan	3400	1020	3.33	NS	NS
Tamil Nadu	6689	2325	2.88	NS	NS
Uttar Pradesh	3875	2320	1.67	+S	NS
West Bengal	5654	3197	1.77	+S	NS

Source: Vaidyanathan et al., 1994.

During 1979–83, taking all crops together, one hectare of irrigated crop area in the states covered by the estimates yielded, on average, Rs 4900, the figure ranging from Rs 3000 in Bihar to Rs 7400 in Maharashtra. Irrigated agriculture in the south Indian states is relatively more productive in comparison to the average; and relatively less so in the east Indian states of Bihar, Orissa, and Madya Pradesh.

Production per irrigated crop hectare is about Rs 2900 (or 50 per cent) more than a hectare of unirrigated area in the selected States taken as a whole. The difference is again the lowest in Bihar (about Rs 700 or 30 per cent of the level of unirrigated crops) and highest in Maharashtra (Rs 5800 or about three and a half times the unirrigated yield).

Since cropping intensities differ substantially between the two categories of land, it is more meaningful to compare their production per unit of land (rather than cropped area). This is estimated on the assumption that gross irrigated crop area is equal to gross cropped area on irrigated land and that gross unirrigated crop area equals the gross crop area on unirrigated land. Cropping intensitites being invariably greater than unity, this measure gives a higher figure of productivity for both categories of land.

The procedure however underestimates the productivity difference between irrigated and rain-fed crop area. First, it has been assumed that area under crops grown largely in irrigated (rain-fed) conditions are wholly irrigated (rain-fed); and also that the overall average yield of such crops represent the irrigated (rain-fed) yield. Irrigated yields being invariably higher, the latter assumption clearly understates the yield of the irrigated component of such crops even as it overstates that of the rainfed component. Second, the entire area under irrigated crops is not covered by the estimates and the crops excluded (vegetables, condiments, spices, flowers) are often those of high value. Third, irrigated areas sometimes grow superior, higher value varieties of particular crops in comparison to those grown on rain-fed land, rice and cotton being good examples.

The extent of underestimation in yield per unit of irrigated land (and upward bias in relation to rain-fed land) is even greater because of the assumptions regarding crop intensities. While, by definition, a piece of unirrigated land cannot grow any irrigated crop, a piece of irrigated land can grow an irrigated crop in one season and a rain-fed one in another. Because of this, the ratio of gross to net irrigated areas tends to underestimate the cropping intensity on irrigated land even as that of gross to net unirrigated area exaggerates the intensity with which rain-fed land is cultivated. In sum, therefore, our assumptions and procedures overestimate the productivity of rain-fed land and understate that of irrigated land as well as the difference between the two. Since the magnitude of the bias varies between region, the estimates of relative productivity differentials across States need to be interpreted with caution.

Time series estimates of the productivity of irrigated and rain-fed crops, using the same assumptions and procedures, have been made for selected States covering the 1970s and 1980s. Trend analysis of these series indicates that, in a large majority of States, irrigated yields show a statistically significant rising trend over this period even as production per unit of unirrigated area does not exhibit a statistically significant trend in either direction. Productivity of unirrigated land is thus more or less stagnant and the productivity differential between irrigated and rain-fed areas has progressively widened both in absolute and relative terms. Moreover, irrigated yields are generally more stable than rain-fed yields, (Rao et al., 1988; Dhawan, 1991; Vaidyanathan et al., 1994).

The stagnation of rain-fed crop productivity in most regions does not square with independent evidence of the development and diffusion of better seed varieties for at least some dryland crops and, more importantly, of the progressive diffusion and rising levels of fertilizer use in rain-fed areas. If true, the implications (namely stagnation, if not decline, in the income of the large segment of agricultural population dependent on rain-fed agriculture and the widening disparities between them and those having the benefits of irrigation) are serious.

Table 10, which compares the average output per hectare across districts grouped according to level of rainfall and irrigation, clearly shows that in a given rainfall zone, districts with higher levels of irrigation use more fertilizers per hectare and produce higher value of output per hectare. This broad pattern has not changed during the last three or four decades.

The level of irrigation development and its expansion also has a bearing on the growth of output Table 11 which gives the mean values of irrigation, fertilizer use, and changes in this for districts grouped by the overall growth of output, shows that faster growth goes with higher irrigation ratios, and large increases in the irrigation ratio. The synergy between irrigation and fertilizers is underlined by the fact that a higher level of irrigation is associated with high efficiency of fertilizer use.

Attempts at applying multiple regression analysis to assess the relative contributions of rainfall, irrigation, irrigation quality, and fertilizers in explaining spatial and temporal variations in productivity have not been particularly successful or conclusive (see Vaidyanathan, 1980; Mukherjee and Vaidyanathan, 1988; Vaidyanathan et al., 1996). The estimated coefficients do not show a pattern and, not infrequently, go against apriori expectation. Part of the reason may be the deficiencies in data. More importantly, however, multiple regression models used for the purpose implicitly assume linear additive relations between inputs and output, when in reality input–output relations in agriculture are complex, non-linear, and interactive in character. Given the formidable problems of specification, estimation, and interpretation, such exercises are of limited value in terms of the insights they have to offer. Direct surveys of the experience of different types of irrigation works under different agro-climatic conditions are essential to obtain a clearer, more accurate picture.

TABLE 10

Mean Values of Selected Charateristics for Districts Classified by Normal Rainfall and 1962–65 Irrigation Ratio

Zone	1962–65				1970–73				1980–83			
	Out-put per ha.	Ferti-lizers*	Irri. ratio@	Rain-fall#	Out-put per ha.	Ferti-lizers*	Irri. ratio@	Rain-fall#	Out-put per ha.	Ferti-lizers*	Irri. ratio@	Rain-fall#
LR LI	409	1.90	7.10	572	507	9.50	10.60	504	664	20.80	20.10	562
LR HI	826	5.30	38.50	625	1154	26.80	49.60	575	1487	63.30	57.10	599
MR LI	564	1.70	7.30	892	608	10.00	10.30	817	780	23.30	17.50	905
MR HI	890	5.30	34.70	910	1055	24.60	40.40	868	1377	61.80	48.50	941
HR LI	758	1.70	5.40	1515	797	10.90	9.50	1464	1004	25.50	14.10	1502
HR HI	1080	6.00	38.20	1354	1268	28.90	41.60	1299	1513	58.00	46.10	1409

* In kg of nutrients/ha. (N.P.K)
@ Irrigation ratio is the ratio of gross irrigated area under all crops to the total gross cropped area (GCA). I have taken the total area under 41 crops as a close approximation of GCA. Estimates of fertilizer use per ha. are also computed from the data on total nutrients consumed and the total area under 41 specified crops for each district.
# in mm.
LR – Low Rainfall (< 750 mm)
LI – Below Average Irrigation Ratio
MR – Medium Rainfall (750 mm–1050 mm)
HI – Above Average Irrigation Ratio
HR – High Rainfall (>1050 mm)
Source: Reconstructed from Bhalla and Tyagi, 1989.

TABLE 11
Per Hectare Value of Output and Input Use and Percentage Gross Irrigated Area in Districts Classified by Growth Rate of Outputs

Output growth rate*	No. of dists.	Value of Output Dists. (Rs./ha.)		Fertilizer consumption (kg/ha.)		Tractors (per 000 ha.)		Pumpsets (per 000 ha.)		Percentage gross irrigated area		Incremented output to fertilizer ratio
		60s	80s	60s	80s	60s	80s	60s	80s	60s	80s	80s
>5.00	24	793.06	1786.80	4.91	94.12	1.71	7.87	7.78	56.00	47.65	79.14	11.10
3.50–5.00	36	745.53	1299.65	4.90	53.58	0.61	3.04	6.01	35.54	33.54	56.77	11.40
2.50–3.50	54	687.48	1074.12	4.94	51.83	0.36	1.63	7.24	32.70	20.44	34.97	8.30
1.50–2.50	65	658.18	877.85	3.50	32.33	0.28	1.23	4.84	26.50	17.88	27.91	7.60
0.00–1.50	77	826.71	922.74	4.00	29.36	0.22	0.80	8.41	35.32	21.59	28.51	3.80
<0.00	25	756.00	749.40	3.40	29.83	0.34	0.66	10.86	59.24	21.51	28.30	–ve
All Districts	281	743.38	1065.25	4.22	44.13	0.43	2.03	7.28	36.59	23.98	38.30	8.10

Note: 60's = 1962–65
80's = 1980–83
* Average annual growth rate between 1962–65 and 1980–83.

Source: Bhalla and Tyagi, 1989.

Direct Survey Based Estimates

Several micro surveys have sought to compare the characteristics of agriculture in areas commanded by irrigation works and those outside (and in a few cases between irrigated and unirrigated lands in the same area). These generally show that irrigated land is used more intensively (both in terms of the extent of fallowing and of multiple cropping); that irrigated land grows more of higher yielding and high priced crops; and that average yields of practically all crops are substantially higher in irrigated areas. Data compiled from a few selected studies of this type (see Table 12) show the gross value of output per unit of gross sown area (GSA) in the command area to be between 7 per cent and 340 per cent

TABLE 12
Impact of Irrigation on Agricultural Productivity

Project	State		Comparision of areas within and outside the command of selected surface irrigation projects		
			Percentage irrigated	GCA/NSA	Output per unit GCA (Rs)
1. Cuttack	Orissa	Command	NA	159.00	3640.00
		Control	NA	129.00	1030.00
2. Mahi II	Gujarat	Command	65.00	136.00	1189.00
		Control	34.00	100.00	808.00
3. Dantimada	Gujarat	Command	63	159.00	2172.00
		Control	58.00	126.00	2017.00
4. Godavari Pravara canal	Maharashtra	Command	57.00	112.00	70.00
		Control	5.00	102.00	16.00
5. Tribeni canal	Bihar	Command	83.00	130.00	563.00
		Control	31.00	111.00	242.00
6. Cauvery-Mettur	Tamil Nadu	Command		129.00	328.00
		Control		100.00	81.00

Source:
1. Satpathy (1984)
2. Adhvaryu et al. (1980)
3. Adhvaryu et al. (1983)
4. Gadgil (1948)
5. Divakar Jha (1967)
6. Sonachalam (1963)
NA = Not Available

higher than in the control area; the difference in productivity per unit of net sown area (NSA) is invariably larger and ranges from 35 to 400 per cent.

Comparison of productivity between irrigated and unirrigated lands in the same area would be a better basis to assess the impact of irrigation, but such surveys are rare. The few available studies do show that irrigation makes a greater difference to land productivity than the control–command area comparisons would suggest. For example, in the Dantiwada project of Gujarat (Adhvaryu et al., 1983) the output per hectare of gross crop area in the command area is estimated to be only 5 to 10 per cent higher than in the control area, but within the command area, the irrigated land is estimated to produce nearly twice the per hectare of gross crop area in comparison to unirrigated land. Another survey in Karnataka reported even larger differences ranging from five to over twentyfold (Nadkarni et al., 1979). The micro surveys also show that the productivity impact of irrigation varies between projects and across regions.

Recently Dhawan (1989) has reviewed a number of other surveys of irrigated areas and attempted to assess the impact of irrigation under different situations. These include comparisons of per hectare production in command areas of ten canal systems from different parts of the country with that of contiguous rain-fed areas; comparisons between areas using groundwater in conjunction with surface water, solely canal irrigated and solely well irrigated farms; and effect of surface and groundwater irrigation in high rainfall areas. The overall productivity impact of irrigation is significant but highly variable across regions, irrigation categories and class of farmers. Other noteworthy findings are that productivity under conjunctive use is not always higher than with canal irrigation alone; that in Maharashtra, the output per unit of water (expressed in rice equivalents) is much higher for water-light crops like coarse grains than for water-intensive crops like sugar-cane; and that returns to groundwater irrigation in east India are modest compared to the western Gangetic plain and favourable to large farmers in both cases. Being relatively few, and not always comparable, these surveys are of limited use in exploring the factors underlying these variations. They do nevertheless suggest that climatic factors and the nature of the irrigation source to be important. For instance, data relating to command areas of some 20 different medium-sized projects in Gujarat (Adhvaryu and Patel, 1984) show cropping intensity as well as crop productivity per hectare are positively correlated to the irrigation ratio. At the same time, projects in the Gujarat region (with relatively higher rainfall) show a higher cropping intensity at comparable levels of irrigation than those in the relatively drier Saurashtra. There is a more pronounced and sustained relation between the irrigation ratio and per hectare yields in Saurashtra than in the western Gujarat region, indicating that the impact of irrigation is greater in the dry region.

The source of irrigation makes a difference to productivity: in Karnataka the areas irrigated by wells have a higher cropping intensity and yields than tank-fed areas (Nadkarni et al., 1979). Surveys of farms served by varying combinations of canals, traditional irrigation sources, and state and private tubewells in Uttar Pradesh suggest that a combination of canals and private tubewells generally does better than others in almost all respects, namely, crop intensity, crop pattern, and yields (Moorti, 1976; Pant, 1984). Among surface sources, the impact of tanks and other local water harvesting systems on the level of yields and their variability relative to rain-fed land is likely to be much lower than reservoir based systems because the latter generally provide larger, more assured, supplies. However there are wide variations between surface projects in terms of quantum, seasonal distribution, and reliability of supplies as well as in their management. These variations and their impact on land productivity have not been systematically studied.

Suggestive as these are, they are an inadequate basis for identification of any generalizable patterns. Systematic collation and analysis of data from many more such surveys could no doubt improve our knowledge of the impact of irrigation. To be meaningful, however, the surveys must be comparable in terms of objectives, design, and survey procedures. Also, given the considerable year-to-year fluctuations and changes in the quality of irrigation and agricultural technology, productivity comparisons of irrigated and unirrigated areas in a particular year can be misleading. Comparisons should be based on at least three year averages before and after a project goes into operation and at periodic intervals for 15–20 years thereafter. Periodic sample surveys of farms in the command areas of select irrigation projects of different types and of wholly rain-fed areas in the neighbourhood repeated periodically (maintaining comparability of concepts and method) is the best way of getting a reliable picture of the dynamics of irrigation and its impact. Until this is done, we have to rely on such data as are available, making imaginative but cautious use of them.

DISTRIBUTION OF IRRIGATION BENEFITS

Irrigation facilities, like land, are unevenly distributed as between regions and between different classes of cultivators. The irrigation ratio in 1988–90 ranged from a mere 11 per cent in Maharashtra to over 90 per cent in

Punjab. As between farms of different sizes, the inequality in the distribution of irrigated land was traditionally less than that of total cultivated land. In other words, smaller holdings tended to have a higher irrigation ratio than the large ones. This has been the pattern in most States. There is basis to believe that inter-State inequalities in access to irrigation facilities have been progressively reduced. There is also evidence of a weakening in the traditional inverse relation between the size of the holding and the irrigation ratio.

The coefficient of variation (CV) across major States (Table 13) in the ratio of net irrigated to net sown area has fallen from 77 per cent in the early sixties to 54 per cent in the late eighties. This reduction in regional disparities appears to be largely due to a more even spread of canal irrigation. Though the CV of the area irrigated by groundwater relative to net sown area has fallen, the fall in CV of surface irrigation areas as proportion of net sown area is much more pronounced. Public investment has thus made a significant contribution to narrowing spatial disparities in irrigation.

As for the impact of irrigation development on different classes of cultivators, there is no strong reason to suppose a significant or universal large farmer bias in the case of surface irrigation works. This bias could however be quite pronounced in the case of groundwater. Wells, energized pump-sets, and tubewells are for the most part in private hands. The outlays required to set these up are typically beyond the means of the smaller cultivators. Though the state and public financial institutions provide liberal loan assistance, such assistance is widely believed to be availed of largely by large farmers partly because the investments are not remunerative unless a certain minimum area can be irrigated.

In principle, small farmers could overcome these problems through cooperative ownership by groups of farmers, and/or the emergence of market for groundwater. By all accounts, joint ownership is limited and largely confined to close kinship groups. Groundwater markets have emerged and grown in all parts of the country, but very unevenly. In some areas they are extensive and function more or less competitively; and at the other extreme there are areas where hardly any groundwater markets exist. This appears to be related to differences in geological conditions that determine the supply of exploitable groundwater and the magnitude of availability in relation to the area in need of it. In general, larger farmers seem to be the principal source of sale of well water to others (Shah, 1993; Janakarajan and Vaidyanathan, 1997). These factors, taken together with the rapid increase in the relative importance of groundwater as a source of irrigation, are the basis for the apprehension that large farmers

TABLE 13

Irrigated Area Total and by Sources Relative to Cultivated Area, Major States 1961/63–1988/90

(percentage)

States	1961–63 Net area irrigated Surface	Ground-water	Total	Gross	1971–73 Net area irrigated Surface	Ground-water	Total	Gross	1981–83 Net area irrigated Surface	Ground-water	Total	Gross	1988–90 Net area irrigated Surface	Ground-water	Total
Andhra Pradesh	11.24	3.58	27.08	29.21	13.05	5.41	27.17	30.14	15.79	7.43	30.92	33.75	16.42	10.22	36.56
Assam	17.18	0.00	28.86	22.83	0.00	0.00	23.83	19.10	13.38	0.00	21.14	16.14	13.38	0.00	21.14
Bihar	7.14	3.86	23.57	19.89	10.90	7.97	28.26	25.99	13.46	12.21	34.12	34.60	15.07	17.42	43.03
Gujarat	1.05	5.92	7.45	7.43	2.28	11.42	14.25	14.85	4.64	17.96	23.09	24.08	4.81	20.96	26.13
Jammu & Kashmir	39.54	0.25	40.80	36.23	40.22	0.32	42.02	38.59	41.12	0.43	43.61	40.59	39.75	0.38	42.81
Karnataka	2.60	1.54	9.04	9.36	4.44	3.15	12.27	13.46	1.01	0.10	14.50	16.36	7.67	5.93	18.90
Kerala	7.73	0.10	16.92	19.74	10.22	0.25	20.38	20.94	4.85	0.47	11.72	13.54	4.82	2.55	14.32
Madhya Pradesh	2.88	2.09	6.09	5.45	4.06	3.49	8.98	8.27	6.05	5.80	13.75	12.19	7.27	8.34	18.38
Maharashtra	1.38	3.49	6.19	6.69	1.67	4.52	8.00	8.89	4.84	6.30	10.79	12.37	5.02	6.65	11.31
Orissa	3.83	0.65	16.59	16.67	9.72	2.33	15.24	17.16	16.72	4.69	28.35	22.40	14.67	10.08	28.81
Punjab*	56.10	26.41	83.98	43.58	30.15	29.16	59.62	62.20	33.58	39.32	73.15	77.22	35.63	46.99	83.02
Rajasthan	4.47	7.68	13.50	12.51	5.36	8.08	14.80	15.37	6.51	12.07	19.80	21.16	8.19	14.91	24.15
Tamil Nadu	14.86	9.97	40.94	44.92	14.83	13.99	44.24	47.31	15.02	17.37	45.01	47.30	13.66	19.59	43.03
Uttar Pradesh	11.93	15.19	31.11	25.87	14.35	23.78	41.64	36.39	19.04	34.81	56.59	48.15	17.97	38.72	59.22
West Bengal	16.51	0.29	26.20	22.69	0.00	0.00	27.26	21.18	12.32	10.79	33.38	24.96	13.44	13.33	35.82
All India	7.93	5.58	18.72	18.81	9.34	9.19	22.88	23.78	11.46	13.52	28.98	29.69	11.98	16.29	32.31
CV	112.2	129.3	76.5	57.1	101.3	112.0	56.4	57.8	102.6	76.7	56.1	58.3	69.3	89.3	54.1

Note: (i) Net = Net irrigated area / Net sown area × 100

(ii) Gross = Gross irrigated area / Gross cropped area × 100

(iii) Surface Source = Canal + Tanks + Other Sources

(iv) Groundwater = Tubewells + Other wells

(v) * Punjab + Haryana

Source: GOI, Ministry of Agricultural Statistics, Various Issues.

may have cornered a disproportionate share of the increments of irrigated areas.

The National Sample Survey (NSS) data on landholdings, which permit a comparison of the changes in irrigation ratio in different size classes of holding between 1953 and 1991 suggests (Table 14) a more complex picture: historically, small farmers have had a higher proportion of their land under irrigation than the large ones. In fact, there has been a clear inverse relation between holding size and irrigation ratio overall as well as under surface and groundwater. That the expansion of both sources during the post-Independence period has benefited all classes of landholdings is evident from Table 14. All six classes also report a progressive rise in the proportion of irrigated area using groundwater. However, the larger farms have experienced a much larger increase in the irrigation ratio by both sources than small holdings. (This tendency is more pronounced in the case of surface water irrigation than groundwater.) Though small farmers continue to have a higher irrigation ratio than average, the differential has progressively narrowed.

Comparisons of trends in irrigation across size classes of holdings at the national level should however be interpreted with caution. There are great regional variations in holdings size, and as well as the level, pace, and nature of irrigation development. The all India distribution reflects the combined effect of inter-regional and inter-class differences in the pattern of irrigation development. Therefore a more disaggregated study of the inter-class distribution of irrigation facilities by States and regions is necessary for a more definitive assessment of the inter-class distribution of irrigation benefits.

A third, and related, aspect that has a bearing on distribution has to do with the way irrigation water is utilized. In any given system, in principle, the available supplies can be used to sustain farming at varying levels of intensity involving different degrees of multiple cropping, crop patterns (especially in regard to the proportion of area devoted to water intensive crops), and application of biochemical technology. In general, a higher intensity of cropping is expected to result in a higher level of output per unit of irrigated land. At the same time, the area irrigated by a given supply tends to be smaller when it is used for intensive cropping than under a more extensive pattern. Intensive irrigation therefore tends to concentrate the benefits by way of additional output and incomes to a relatively smaller area and, by implication, to fewer farmers when compared to 'extensive' irrigation.

The distribution of available water to a larger number of farmers for

TABLE 14

Irrigated Area by Sources as Percentage of Total operated Area by Size Class

Size Class	1953–54			1981–82			1991–92			Percentage change in irrigated ratio 1953–91		
	Surface water	Ground-water	Total	Surface water	Ground-water	Total	Surface water	Ground-water	Total	Surface water	Ground-water	Total
<1 ha.	18.52	8.04	26.56	19.84	19.05	38.89	19.76	21.84	41.60	6.70	171.64	56.63
1–2 ha.	18.02	7.13	25.15	17.64	18.03	35.67	18.07	20.83	38.90	0.28	192.15	54.67
2–4 ha.	14.28	6.76	21.04	13.71	18.48	32.19	15.89	20.91	36.80	11.27	209.32	74.90
4–10 ha.	9.80	5.36	15.16	11.45	17.70	29.14	13.85	19.02	32.80	41.33	254.85	116.36
>10 ha.	4.39	2.43	6.82	6.30	8.78	15.07	15.55	12.55	28.10	254.21	416.46	312.02
All size class	9.95	4.87	14.82	13.19	16.54	29.73	15.97	19.27	35.40	60.50	295.69	138.87
CV	40.9	32.9	38.2	34.5	23.4	27.3	16.6	17.70	14.40			
As per land use statistics*	11.97	5.27	17.25	15.21	12.75	27.96	12.93	18.66	31.59	8.02	254.08	83.13

Source: GOI, NSSO, No.74, *Sarvekshna*, July 1988 and 48th Round, March, 1997.

*GOI, Ministry of Agriculture, *Indian Agricultural Statistics.*

Note: NSS figures relate to the ratio of net irrigated area to total operated area, which is larger than net sown area. LUS relates refer to ratio of net irrigated area to net sown area.

Groundwater = tubewell + wells

Surface water = Canal + Tanks + Others + not recorded.

growing relatively less water intensive crops makes for a wider diffusion of benefits to a larger number of people. From a social viewpoint, this is clearly preferable if it also leads to at least as large an increment to output per unit of water as the intensive pattern. Production per unit of water is an important consideration in situations where irrigation supplies cannot meet the full requirement of the entire area under cultivation and is a limiting factor on the use of other yield raising inputs as well as their effectiveness.

Those who argue for intensive irrigation appear to assume that it invariably produces more output per unit of water consumed. It is however important to distinguish between two different aspects of the notion of 'intensive irrigation'. One has to do with ensuring that a given crop gets ample, assured, and timely supplies to maintain the soil moisture regime conducive to effective exploitation of the potential of available biochemical technology for that crop. The second has to do essentially with the kind of crops grown and the number of times a year a piece of land is cropped (or to be more precise, the number of months a year the crops are grown on a given piece of land).

Though firm quantitative evidence is hard to come by, there is good reason to believe that the soil moisture regime does affect yield response to fertilizers, and that this is particularly pronounced in the case of the HYVs. Scattered farm surveys suggest that private tubewells, or a combination of canal and tubewell irrigation, goes with larger water deliveries per unit area, higher input intensity, greater use of HYV, and high yields (Moorti, 1976; Pant, 1984). Evidence on the value of productivity of different crops per unit of irrigation water is, however, scarce. That there is a widespread tendency for the farmers to prefer water intensive crop patterns, and to adopt them in the face of legal restrictions and risking heavy penalities, is a clear pointer to their superiority in terms of private returns. However, this does not necessarily mean that they are also socially preferable in terms of the contribution to output per unit of water used.

Data from Maharasthra in fact suggest that contrary to belief, sugar-cane, which requires at least eight times as much irrigation water as seasonal crops like jowar and bajra, and six times that of other 'seasonal crops', is estimated to generate among the lowest net farm business incomes, per million cubic feet of water consumed. It has been argued that 'if any of the many seasonal crop combinations/rotations are adopted in place of sugar-cane, not only will the net total income from the given quantum of water be higher; it will cover a wider net irrigated area as well' (Rath and Mitra, 1987). On this basis, a strong case has been made for restricting the year round supply of public canal water for sugar-cane.

Whether this would be generally valid for all regions is difficult to judge. The feasible crop patterns, the irrigation needs of crops and their yield performance depend not only on the irrigation regime but also on the climatic conditions (which vary from region to region) and the state of biochemical technology. Judgements on these questions become more complicated when public canals are used in conjunction with private wells which themselves derive part or all their supplies from canal seepage. The optimum solution also depends on costs associated with water control systems of varying coverage and sophistication.

PROBLEMS AND RESPONSES

That the development of irrigation facilities has made a major contribution to sustaining agricultural growth during the past five decades is beyond question. However, experience has thrown up a number of problems, and irrigation programmes and policies have attracted criticism on several grounds. Inordinate delays in completion of large surface projects and underutilization of their potential, unsatisfactory quality of irrigation, low cost recovery, the adverse ecological and social consequences of the present pattern of development are among the principal issues that figure prominently in both official documents and public discussions. It is useful to review these problems, the manner in which they have been addressed, and the lessons that can be learnt regarding more effective ways of dealing with them.

Delays and Underutilization

Large cost overruns and delays in the execution of projects have been long-standing complaints about Indian irrigation planning. There are relatively few detailed and comprehensive studies of this phenomenon.[5] One such study which examined 30 completed projects found the actual cost to be more than twice that of the original estimate. The estimated average time taken for completing projects (based on a study of 8 major projects) was nearly 11 years, about 4.5 years more than twice the original estimates.

It is a measure of delays in completing project construction that, at the end of the Sixth Plan, the actual expenditure on projects started before 1974 was barely 40 per cent of the revised cost estimates; and the actual potential created one-fourth of the ultimate potential expected. As against the expectation that all the 65 major projects started before 1976 would be completed by the end of the Sixth Plan, in actual fact, only 25 were completed. At the end of the Seventh Plan out of the 182 major ongoing projects, the proportion of actual outlay to estimated cost exceeded 75 per cent only in 56; the outlays incurred were less than 75 per cent of the estimated total cost in about two-thirds of both major and medium projects; less than 30 per cent of the cost has been incurred in nearly a third of the major projects and a fourth of medium schemes (GOI, PC, 1992: 88).

The factors contributing to cost escalations and the long gestation are known in a general way. They include inflation in the prices of the inputs that go into the construction of irrigation projects; design modifications due to inadequate preparatory investigations; significant changes in the scope of projects after they are approved; starting far too many projects in relation to the available budget and resources; procedural delays in land acquisition, bottlenecks in material supply.

However, there is no study of the contribution of these elements in accounting for revisions in cost estimates either overall or for specific projects. Recently the CWC published an index of the construction costs of irrigation projects reflecting the rise in prices of inputs (GOI, CWC, 1991).[6] The deflated figures of overall investment in relation to the potential created suggests an increase in average real outlays per hectare. This is however not a reliable basis on which to judge the trend in real costs because they include a large element of outlays on incomplete projects. In any case they tell us little about the causes of the changes in cost.

In theory, projects are supposed to be given investment clearance only after ensuring that they are based on adequate investigations, that the technical designs and cost estimates are sound, and that the project is economically viable. Revisions in cost estimates exceeding 10 per cent are also subject to review and approval by the Technical Advisory Committee (TAC) of the Central Water Resources Ministry. The experience of the last 40 years shows that in reality the standards of scrutiny and evaluation are lax, that project proposals without adequate hydrological data and preparatory investigation have been approved under political pressure, that many projects have been initiated by States without any approval by the TAC, and that changes in the scope, design, and cost estimates are large and frequent. It is surprising but true that there are no rigorous and detailed analyses of the factors responsible for revision in cost estimates or the justification for the large divergence between projected and actual costs.

Successive plans draw attention to the need to check delays and cost escalations. The latest in the series, the Eighth Five Year Plan, states:

It has become important to tighten the standards of projects preparation and design.... In particular there is need to pay much greater attention at the planning stage to systematic surveys of soil conditions, land use capability, crop water needs, irrigation efficiency and conjunctive use of surface and groundwater to ensure that the command area, the distribution network and crop patterns are based on firm data, and are mutually consistent.... The project proposals submitted for approval must also examine various options for use of water in a project and choose the optimal alternative keeping both efficiency and equity in view. [GOI, 1992: 59.]

It goes on to assert that 'completion of ongoing projects with a strict prioritization will be the first charge on available funds,' and that 'in the case of ongoing projects in early stages of construction and in the case of new projects the state governments will be required to prepare... a detailed programme for OFD works and water management, plan based on detailed soil surveys and land use capability' (ibid.).

All this is unexceptionable. But the experience of the past 40 years leaves no room for optimism that the days of permissiveness and mindless pursuit of shortsighted competitive politics that currently shape investment decisions are over. Political pressures for starting new irrigation projects are very strong. Since such projects stand to benefit a large number of farmers, politicians in power view them as instruments for mobilizing electoral support. Large construction projects also happen to be a convenient means for mobilizing funds for party and other uses. The frequent turnover of both ministers and bureaucratic managers responsible for irrigation further shortens the time horizons of decision-making. Since the consequences of bad investment decisions or delayed implementation rarely visit the individual ministers/officials who make the decisions, there is little check on several undesirable tendencies: namely, (a) the propensity to commit the government to new projects without the necessary technical preparation and unmindful of the economic viability of investments; (b) to exert political pressure to get engineers to prepare projects in a hurry and obtain the necessary clearance; and (c) to start far more projects than can be accommodated within the available resources. Unless these tendencies are curbed, strict enforcement of standards of design, efficiency and cost control in projects will remain elusive.

Utilization of Potential

As has already been noted, current perceptions of the problem of underutilization of potential created are oversimplified and may well be misplaced. Nevertheless, since the early 1970s, it has been considered sufficiently important to merit special attention. The switch from rain-fed agriculture to irrigated agriculture involves a major change in agricultural techniques that farmers take time to master. The Irrigation Commission

(GOI, 1972) which went into the problem at some length, argued that utilization is impeded by deficiencies in the design and implementation of projects, including completion of reservoirs ahead of the canal networks, lack of field channels to carry water from the government outlets to the individual farms, inadequate attention given to proper preparation to make land fit for irrigated farming, and poor planning of cropping patterns suited to the varying soil and drainage conditions in different parts of the command area.

Based on this diagnosis, the Commission suggested the creation of a unified organization in the command areas of major projects to achieve speedier utilization of irrigation potential and ensure proper use of water. Accordingly, the Command Area Development Programme (CADP) was started in 1974–5. It

envisaged execution of on-farm development works (like field channels, land levelling, field drains and conjuctive use of ground and surface water); the introduction of rotational system of water distribution to ensure equitable and timely supply of water to each and every farm holding; and evolving and propagating crop patterns and water management practices appropriate to each command area. Other ancillary activities like construction of link roads, godowns and market centres, arrangements for supply of inputs and credits, agricultural extension and development of groundwater for conjunctive use are also taken up as part of the relevant sectoral programmes in the State Plan.

The programme was initially initiated in 60 major and medium projects. Gradually more projects were brought into its ambit, and it currently covers 131 irrigation projects with a total command area of around 18.5 m. ha. Up to the end of the Seventh Plan Rs 22 bn. have been spent on the programme.

In actual practice the scope of the programme proved out to be much narrower than envisaged, focusing largely on the construction of field channels, introduction of *warabandi* and land levelling. Achievements in these areas have also been somewhat modest. At the end of the Seventh Plan an estimated 8 m. ha. had been brought under *warabandi*; 2 m. ha. had been levelled and shaped; and field channels constructed for 11.3 m. ha. Hardly any progress had been made in land consolidation. In all these areas the achievements have been consistently below targets (Table 15).

On the whole, as the Eighth Plan noted, 'The progress in terms of land improvements and improvement of drainage facilities has been meagre, and so has the effort and research in evolving and propagating cropping patterns and agricultural practices for optimum use of water

TABLE 15

Targets and Achievements in CAD Works

(million hectare)

	Fifth Plan ach.	Sixth Plan		Seventh Plan	
		Targets	Ach.	Targets	Ach.
Construction of field channels	2.80	4.00	5.23	6.81	3.26
Warabandi	Not introduced	Not Fixed	1.66	8.03	4.34
Land levelling	1.10	1.00	0.50	1.82	0.42

Source: GOI, GC, Eighth Five Year Plan.

under the conditions prevailing in each irrigation command' (GOI, PC, 1992). Organizational structures for promoting the proper use of water have not been put in place. Only in a few cases have the command area authorities been given control over the management of the system as a whole. In most cases the CADA is only responsible for carrying out physical works and occasionally management of water below the outlets. The system of secondment of officers from different departments has not been conducive to the development of an integrated management system.

There are few evaluations of the impact of CAD on the extent of utilization and adoption of desired cropping patterns. The most comprehensive studies available relate principally to the organizational aspects and the constraints on the implementation of certain categories of physical improvements in the system. Very few CADPs conduct regular crop cutting experiments or collect reliable data to assess the impact on production and productivity.

The rationale for special command area development clearly needs re-examination in the light of the experience. The various measures by which CADP seek to ensure speedy and efficient use of water from surface systems must in any case form an integral part of project planning and management, and not as a separate activity to be taken up after a project is commissioned. Nor is it prudent for the government to assume responsibility for field channels, land improvement, and other on-farm works. By doing so the stake of the beneficiaries and their involvement in the proper maintenance and upkeep of the facilities, already weak, is further eroded.

Quality of Irrigation

Far more important is the quality of irrigation reflected in the volume and duration of water available per unit of irrigated area; the assurance and timeliness of supply, and ability to adjust the supply according the needs of particular crops in particular segments of the irrigated area. Substantial improvements have taken place in all these respects.

Changes in the relative importance of different sources of water supply have been a significant contributing factor: in the case of surface irrigation there has been a massive shift from works that draw their supply from small local storages and streams to larger reservoirs. In 1950–1 canals (government and private) irrigated an estimated 8.3 m. ha. of cultivated land compared to 6.6 m. ha. served by tanks and other minor surface works. Most of the canal irrigated areas were located in the deltaic tracts of east flowing rivers in peninsular India and the Indo-Gangetic plain. They drew water supply by diverting the river flow with the aid of low barrages. Large storage reservoirs were relatively rare: the capacity of all types of storages in 1950–1 is placed at no more than 30 bn. m^3, about half of which was accounted by tanks.

The situation has since changed radically: between 1950 and 1990 the government has made massive investments in developing large-scale, reservoir-based canal irrigation. Over 60 per cent of the total public sector plan outlay on irrigation and flood control has been devoted to this category of works. By comparison, investments in the construction of tanks and other small-scale surface irrigation was miniscule. Consequently, while the storage capacity of tanks increased marginally, if at all, that of modern large reservoirs rose to 163 bn. m^3 at the end of the 1980s, accounting for nearly 90 per cent of total storage capacity. Whatever other disadvantages they may have, large reservoirs harvest water from a much wider catchment and are therefore able to provide a larger quantum of water, for a longer duration and with greater assurance than minor surface works.

A second source of quality improvement is the policy of encouraging conjoint use of ground and surface water in the areas commanded by surface irrigation works. Till the mid-sixties, digging wells in canal commands was either prohibited or subject to severe restrictions. Following the introduction of high yielding varieties, and the realization that ample and assured water supply is essential to exploit their full potential, the restrictions were greatly relaxed in favour of a policy of active encouragement of conjoint use. The principal source of groundwater recharge in canal commands is seepage from surface water application. Conjoint use in effect recycles this seepage, thereby increasing the proportion of surface water supplied which is effectively available for the crops. Being owned and operated by individual farmers on their lands, wells give them much greater control over timing and volume of irrigation. While we have no reliable data on the precise extent of conjoint use, there is little doubt that it has spread rapidly and covers a sizeable part of surface irrigated area.[7]

There has also been a rapid expansion of groundwater used as a sole source of irrigation. The introduction of HYV increased the profitability of irrigation; and government policies—investment in rural electrification, supply of cheap and increasingly subsidized power and credit—have given a powerful stimulus to development of this source. Tubewells, which tap deeper aquifers and provide larger volumes of water than dug wells, have become increasingly important. Overall, according to land use statistics, the extent of land reporting irrigation by groundwater has increased 3.5 to 4 times since 1950–1 in comparison to a 50 per cent rise in area under all surface irrigation sources. Though these data suffer from many frailties, they do indicate a major shift towards better quality irrigation.

The rapid, and practically unbridled, expansion of groundwater use has however been accompanied by a proliferation of the number of wells and also a widespread and substantial lowering of water tables all over the country. The persistence of this trend and its increasing spread across regions has reinforced the well-founded apprehensions of over exploitation of this resource, and about its implications for equitable sharing of benefits. How best to contain the level of groundwater use within sustainable limits has thus emerged as a major and urgent issue.

The spread of conjoint use of groundwater has not benefited all areas served by surface irrigation. The quality of irrigation in areas depending solely or mainly on canals or tanks leaves much to be desired. Most of the minor irrigation sources are relatively old and have been constructed, maintained, and managed by local user communities. Their regulatory structures and distribution systems are relatively simple and far less sophisticated than those used in large systems. Moreover, it is widely believed that these systems have deteriorated because of the weakening of traditional arrangements for managing these facilities, the cumulative effects of neglect of upkeep and repair of damages to structures, and siltation of storages and channels. The government has not done much to address these problems either.

A number of surveys corroborate these apprehensions. However, systematic measurement of their impact on characteristics of water supply and use in the systems are rare. A recent rapid survey of some 100 tanks in Tamil Nadu[8] shows that damage to the inlet channels of tank bunds, sluice regulators, and surplus wiers are quite widespread and often serious. Nonetheless, most the tanks studied are still in use. In most cases, informed members of the community report reduction in water supply and weakening of institutions, but few report any significant reduction in the effective command in normal years. One cannot however confidently assess, on the basis of such limited and rapid surveys, the secular trend in the conditions of water supply from tanks or their impact on irrigated area and productivity. The balance of evidence, however, points to a deterioration. There can be no doubt that their condition offer much scope for improvement.

Major and medium surface systems, though superior to minor surface works in terms of the quality of water supply, face more complex and difficult problems in regulating water deliveries in accordance with the water needs of crops at different times and in different segments of their command areas.[9] These systems are designed on the basis of estimated water flows in the river at a certain level of dependability (usually 75 per cent) and assumptions regarding conveyance and application losses, expected crop patterns in different segments of the command area, and the corresponding irrigation requirements at different times in an average year. The layout and specifications of the distribution network as well as the rules of operation of the reservoir and the canals are derived from these basic features. In theory, the principles and procedures by which shortages in supply are to be managed are expected to be spelt out.

Typically, however, adequate and reliable information on all relevant aspects is not available at the time of design. Moreover, designers are subject to a variety of pressures: for instance, they have to ensure that the project proposal will pass the technical and economic appraisal; and they have to be seen to respond to concerns about distributing the benefits of irrigation as widely as possible. All this can, and does, lead to inconsistencies between water available, command area, crop pattern, and canal/distributary capacities even at the design stage.

The operating schedules and the criteria, rules, and mechanisms underlying them are seldom spelt out explicitly and clearly even for a normal year. The basis for adjusting allocations to cope with situations arising from abnormal seasonal conditions, below-normal water supply, and changing crop patterns are at best fuzzy. Once the system goes into operation these inconsistencies surface; and are compounded by the divergence between the design assumptions on these aspects and what is actually realized. The operational procedures must therefore be somehow adapted to resolve this hiatus. However the adaptation process is seldom based on an explicit re-evaluation of the operational rules in the light of accumulated experience, but happens in a haphazard way by a process of trial and error, in which geography and power play a more decisive role than reasoned assessment from the social viewpoint.

The ability of the system managers to ensure efficient water distribution is constrained by, (a) the rigidity of the distribution network once constructed; (b) the extent and sophistication of control structures to regulate the volume and timing of flow to diffrent parts of the command area; (c) the capacity to collate and analyse the enormous amount of information available bearing on crop water needs at different times in different segments of the command area; and (d) the ability of the system managers to enforce the allocations on the ground. In all these respects, again, the situation in Indian canal systems is far from satisfactory.

The control structures are relatively few and rather crude; the arrangements for collection of relevant data are grossly inadequate and the capacity to analyse them even more so; and the systems, being managed by personnel drawn from the general engineering cadre of the government, does not permit the accumulation of knowledge, skills, and experience in operating such complex systems. In the absence of a clear cut and internally consistent set of operating rules, professional management of such complex systems becomes very difficult, and more so when key operating decisions (opening and closing canals, scheduling water to particular segments) are centralized and open to political influence.

Interference with the functioning of water allocation is universal and takes several forms: arbitrary changes in operating schedules; growing unauthorized crops in unauthorized places; diverting more water from the canal than is permitted digging wells despite restrictions; and unauthorized tapping of water by persons outside the command area. These violations take place by 'persuading' the bureaucracy responsible for running the systems, or by getting centres of power outside the system (e.g. the higher level bureaucracy, the political party network) to intervene to bend the rules and/or prevent enforcement of penalities for violations. Under these circumstances, professional, rule-based management of water allocations becomes virtually impossible (Wade, 1980,1982).

As has already been noted, the Command Area Development Programmes as well as the programme for modernization of older systems seek to address these problems through investment in engineering works to permit better regulation of water and more effective use of water allocations. Some progress has been made with regard to physical improvements, but unevenly. Conjoint use has certainly afforded farmers greater flexibility in water management, but it cannot compensate for shortfalls in canal supplies which are after all the primary source of recharge.

The *warabandi* system which seeks to impose some order and predicability in water supplies presently works best within a season and in areas where the bulk of the irrigated area is devoted to the same crop or to crops with a more or less similar level and pattern of water requirement (as for example, in the *rabi* season in the wheat growing areas). Some studies from the sixties have shown that even in Punjab, where the *Warabandi* system has a long history and is reputed to be effective, the system gives reasonable assurance as to the dates on which the farmers in a particular outlet may expect to get water, but no guarantee as to the amounts they will get. There is in fact considerable variation in the amount of canal water received by the farmers with tail-enders being at a distinct disadvantage (Reidinger, 1974).

In situations where crops with diverse water requirements are grown, the problem is more difficult. The approach so far has been to indicate permissible crop pattern (annual and seasonal; water intensive and water light crops) in different segments of the command area and design the schedules on that basis. But when crop restrictions are not observed and sanctions against violations are ineffective, and this is invariably the case, rational management of water deliveries is practically impossible. It is arguable that a better approach would be to focus only on laying down and enforcing the volume and duration of supply to different segments of the command area, leaving farmers free to decide what crops they want to grow with that level of supply, but this is not a possibility that has been seriously considered so far.

Cost Recovery[10]

Public sector irrigation systems in India run at considerable, and increasing, losses. Users of canal irrigation are charged on the basis of area irrigated. The rates are differentiated between crops: water intensive crops like paddy and sugar-cane being charged more than others. The inter crop differences are not however matched with the differences in their irrigation requirements. In some regions irrigated land served by old systems (including tanks) bears a higher rate of land revenue but does not pay a separate water rate.

The collections on account of water charges have for long been inadequate to cover the costs of operation and maintenance, not to mention capital charges. In 1987–8, irrigation revenues met only a fourth of the total cost of providing the service computed as the sum of operation and maintenance expenses, depreciation at the rate of 1 per cent of cumulated investment, and interest on that investment at the average rate paid on the outstanding debt of the particular State government. Since then, even as all elements of costs have risen, collections have not even proportionally increased. Consequently the cost recovery rate has fallen to an estimated 4 per cent in 1993–4 (Table 16).

TABLE 16

Unrecovered Costs on Account of Major and Medium Irrigation Works, Major States

(Rs m.)

	1987–8[1]	1993–4[2]
Gross revenue	5925	5490
Total cost	52979	129698
Unrecovered costs	47048	124208
Percentage cost (1 as % of 3)	12.60	4.4
Public sector plan outlay on irrigation	33469	58540

Source: [1]Srivatsava et al., 1997.
[2]*Economic Survey*,1995–96.

Low cost recovery is partly due to under-assessment. There is an incentive to under-report both the total area irrigated by canals and the irrigated area sown with crops carrying relatively higher rates. Since the records are maintained by local officials and supervision to ensure their accuracy is inherently difficult and increasingly lax, the opportunity for collusive under-recording is also high. A second reason is that the assessed dues are not fully collected. Available information, which is often incomplete, indicates that accumulated arrears of irrigation charges are large and State governments are reluctant to take the steps necessary (and available under the law) to recover dues.

A third reason is that the general level of rates is far too low. Time was, during the colonial rule, when earning adequate revenue to cover the *full* costs (including capital charges) of providing irrigation was the guiding criteria of government policy. The justification for new investments was also judged by this criterion. The situation however changed after Independence: the provision of irrigation and other infrastructure to facilitate and stimulate economic development came to be accepted as the responsibility of the government. Progressively the populist notion that farmers should be provided water (as indeed other inputs and services) at cheap rates gained ground. It has acquired such momentum and strength that, irrespective of which parties are in power, governments shy away from the problem.

The fiscal consequences of cheap pricing of water have been reiterated by several Finance Commissions, successive Plan documents, as well as numerous expert committees and academic studies. Even moderate suggestions that rates should cover at least the operation and maintenance costs were not acted upon. Failure to recover costs and the resultant 'losses' in effect uses resources to give an indefinite recurring advantage to existing users of public irrigation instead of applying them to expand and improve irrigation facilities to benefit wider areas and a larger number of farmers.

The revenue deficits of most States are mounting even after substantial increases in statutory devolutions of tax revenues from the Centre and rising central assistance. In this context, the large and growing magnitude of unrecovered costs seriously impedes the capacity of States to maintain adequate fresh investment in irrigation and other infrastructure essential to facilitate and sustain a satisfactory tempo of overall economic growth. The seriousness of this problem is underscored by the fact that in mid-eighties, unrecovered costs on irrigation (Rs 47 bn.) amounted to a quarter of the total state Plan outlays and larger (by about 40 per cent) then the state Plan outlays on irrigation. By 1993–4, losses on account of irrigtion had increased to Rs 124 bn., about 2.3 times the Plan outlay on irrigation.[11]

The argument that increases in water charges will hurt poor farmers is palpably wrong: hardly one-fifth of the Indian farmers benefit from canal irrigation and among those that do, the better off farmers receive the bulk of the benefits. On the other hand, the policy of supplying water at prices far below what it costs, greatly weakens the incentives for government agencies to manage facilities economically and for farmers to use scarce water efficiently. The impact of irrigation on productivity is substantial, and there is every justification for the state to appropriate a part of the additional incomes generated by its investment in extending irrigation.

The case for charging users of public irrigation the full cost of supplying water is therefore strong. The problem is to ensure that users are charged only for costs necessary to provide water, and that the irrigtion agencies are obliged to provide a specified standard of service to the users. At present neither of these conditions is satisfied. Capital costs are higher than they need be for several reasons: projects are designed without adequate investigation, procedures for scrutiny and clearance of projects are weak, and the temptation to promise irrigation from particular projects to as large an area as possible is strong. Time and again the decision-makers have readily succumbed to this temptation even though the available supplies can serve only a fraction of the command area. A substantial part of the investment in distribution networks is thus infructuous and wasteful. Corruption in the award and execution of construction contracts add to the costs. The users cannot in fairness be expected to pay for these costs.

The present allocation system is designed and administered wholly by the bureaucracy. Rules of allocation are seldom clearly specified and are liable to be changed at the discretion of the system managers and the government. The entitlement of users are not clearly spelt out, and the mechanisms for enforcing their entitlement are inadequate and ineffective. Irrigation schedules are unpredictable, water supply unreliable, violations of rules concerning crop patterns and illegal appropriation of water are widespread, and the sanctions prescribed by law rarely effective. The operating procedures of canals often conflict with the crop water needs on the ground, thus contributing to disruption of the supplies by those who are favourably placed to do so either because of their advantageous location or power. All this adds up to poor quality of irrigation.

Clearly it is unfair to charge users for the inefficiencies in project construction and management, or without any reference to the quality of service rendered. The appropriate level of rates must therefore be decided after reassessment of the costs that can be reasonably charged to the users, as well as the acceptance on the part of the government to ensure explicitly stated and mutually agreed standards of service to the uses. While an upward revision of water rate is inescapable, it would be unrealistic to expect that the appropriate level of rates can be implemented immediately. The movement to this ideal has to be phased and related to a wider programme of reforms in management of irrigation outlined later in this volume.

Sustainability

Concerns about sustainability arise basically from higher-than-expected rate of siltation of reservoirs, widespread lowering of groundwater tables, and the spread of waterlogging and salinity in irrigated tracts.

Continued deforestation, especially in upper catchments of rivers, and the extension of cutivation to marginal land and ecologically fragile areas have aggravated soil erosion. There is evidence that several important reservoirs are getting silted at a much faster rate than was originally estimated. The effective life of reservoirs, it is feared, will be much shorter than expected. A number of schemes for treatment of catchments of major reservoirs have been implemented, and more are envisaged. Their aim is to arrest deforestation and restore adequate forest cover in the upper catchments of rivers. Their scale is however inadequate in relation to the magnitude of the problem, and the effectiveness of completed schemes in checking siltation remains to be established by proper surveys.

The Eighth Plan indicates that treatment of catchment areas of new irrigated projects is to be made an integral part of the project. Integrated watershed development for soil and moisture conservation of all vulnerable land (forest and non forest), that is now an accepted concept, could make a significant difference, but numerous technical and organizational problems have to be resolved before the concept becomes operational. Programmes commensurate to the needs of the situation will also entail a much larger allocation of resources.

As for groundwater depletion, the central and State groundwater boards have evolved a system of categorizing Blocks according to the scope for further development. The system of classification is not satisfactory because there are no reliable direct measurements by Block of extraction and recharge, and because it is difficult to make meaningful inferences about sustainable rate of exploitation in administrative units whose boundaries do not coincide with aquifer boundaries. Nevertheless, the progressive increase in the number of Blocks where groundwater exploitation is estimated to have reached or exceeded sustainable levels is indicative. Moreover, the large and growing body of evidence on the progressive lowering of the groundwater table is a signal that over-exploitation of this resource is more widespread and serious than has been recognized.

The long-term consequences of a progressive fall in the groundwater table are likely to be different in areas depending solely on well or tube-well irrigation and those where they are used along with surface water. In the latter case, since the volume of the seepage depends on the volume of surface water application, there is no danger to the sustainability of overall supply of irrigation water in the command area. Competitive deepening and proliferation of wells in effect redistributes the recycled seepage between different farmers. In areas irrigated solely by groundwater, lowering water tables will also change distribution of access and benefits, but when unchecked it will eventually result in a disastrous fall in total output and incomes. That is why farsighted measures to contain groundwater use in purely groundwater irrigated tracts are particularly important. State regulation is necessary. Legislation in fact exists but is largely ineffective. While the state has to be 'hard' in enforcing the rules, it is also necessary to create strong disincentives to discourage over exploitation of groundwater. The most effective way to do so is to make it more expensive to pump water.

Waterlogging and Salinity

Injudicious use of canal water, typically arising from over-irrigation and neglect of drainage, causes waterlogging and a rise in the water table which, left uncorrected, eventually lead to salinization. In many places obstruction of natural drainage by the construction of roads, railways, embankments, and the like has disturbed the surface hydrology and aggravated drainage problems. Although irrigation and drainage should both go hand in hand, the drainage aspect has not been given due attention even in the major and medium irrigation projects.

There are a number of differing estimates on the extent of area affected by waterlogging (see Table 17). The Ministry of Agriculture has been of the view that the area affected by waterlogging increased between 1972 and 1990. In 1972, according to estimates of the Irrigation Commission, States like Andhra Pradesh, Assam, Bihar, Gujarat, Kerala, Orissa, and Tamil Nadu were not affected by this problem. The total area affected was placed at 4.8 m. ha. The latest Ministry of Agriculture estimates (relating to 1990) places the affected area at about 8.5 m. ha. and shows significant areas to be suffering from waterlogging in the above-mentioned

TABLE 17
Estimates of the Extent of Waterlogging in India

	Irrigation Commission (1972)	National Commission on Agriculture (1976)	Central Water Commission (1990)	Ministry of Agriculture (1990)
Andhra Pradesh	0.00	0.34	0.27	0.33
Assam	0.00	0.00	0.00	0.45
Bihar	0.00	0.12	0.36	0.71
Gujarat	0.65	0.48	0.23	0.48
Haryana	0.00	0.62	0.00	0.62
Jammu & Kashmir	0.01	0.01	1.00	0.01
Karnataka	0.06	0.05	0.02	0.01
Kerala	0.03	0.11	0.00	0.06
Madhya Pradesh	1.10	0.06	1.00	0.06
Maharashtra	0.35	1.10	0.00	0.11
Orissa	0.00	0.35	4.00	0.06
Punjab	0.81	0.02	0.00	1.10
Rajasthan	1.85	0.81	6.00	0.35
Tamil Nadu		1.85	0.19	0.02
Uttar Pradesh		0.00	0.20	1.98
West Bengal		1.00	0.03	2.18
Delhi			5.00	0.00
Total	4.84	5.98	1.61	8.52

Source: O.P. Singh and I.P. Abrol (1992).

States. The estimates made by the Central Water Commission are much lower at 1.6 m. ha. but they cover only areas waterlogged due to a rise in the groundwater table. The estimates of both the National Commission on Agriculture and the Ministry of Agriculture consider both areas affected by over-irrigation and by rises in groundwater levels as waterlogged.

The spread of conjoint use of groundwater with surface water (especially in Punjab, Haryana, and parts of U.P.) must have substantially lowered the water table and thus helped to contain waterlogging/salinity. There are however apprehensions that the rapid spread of the paddy–wheat rotation in this region will damage the soil, and that the current level of productivity may not be sustainable (Joshi and Tyagi, 1991; Khepar and Sondhi, 1992). Elsewhere there is good reason to believe that the problem is becoming more widespread and acute. Recent studies have indicated that there has been a considerable rise in the water table in a number of irrigation command areas all over the country due to improper water management. According the Ministry of Water Resources, in Andhra Pradesh, the Krishna river delta has experienced a 2 to 4.4 m rise in the water table. In Karnataka, the Chitradurga area has experienced a 2 to 6.8 m increase. In Madhya Pradesh, the rise has been 2 to 9.3 m in certain areas. In Faridkot, Punjab, the rise has been 2 to 11.2 m.

There has been no comprehensive nation-wide survey so far. However, detailed studies have been undertaken about the changes in waterlogging in four irrigation projects: Sriramsagar in Andhra Pradesh, Nagarjunasagar in Andhra Pradesh, Tawa in Madhya Pradesh, and Tungabhadra in Karnataka. Waterlogging and soil salinity are closely related and interlinked. Due to these problems, a substantial part of area has either become unproductive or has gone out of cultivation (Ministry of Water Resources, 1992). According to a 1991 study on the status of waterlogging, salinity, and alkalinity by the Ministry of Water Resources, the problem is widespread in irrigation projects and a matter of serious concern. The areas reported as waterlogged, saline, and alkaline in Table 17 are the newly created problems resulting from the initiation of irrigation projects.

The problem has been long recognized and the principal ingredients of a solution are known. There is, as it were, a preventive and a curative aspect to this. Greater care in providing drainage facilities in canal command areas, planning of crop patterns suited to the elevation, slope, and soil conditions of different segments of the command areas, and getting farmers to avoid indiscriminate water application are preventive steps that should be built into all projects. For areas that are already afflicted we need a systematic survey to assess the extent, nature, and location of the waterlogged and saline/alkaline land in existing project command areas. The cost effective measures to reclaim such affected land and restore it to its full potential have to be decided in the light of nature and intensity of degradation in particular segments. There is however urgent need to go beyond such generalizations to concrete, area-specific action programmes.

FUTURE DIRECTIONS

Given that irrigation makes a big difference to the productivity of land, and that, despite developments over the past five decades, nearly 60 per cent of the cropped area is still rain-fed, it is natural that both farmers and governments are deeply interested in expanding irrigation facilities as rapidly as possible. In the light of past experience, however, several questions concerning to scale and priorities of water resource development merit deeper consideration. How much scope is there for further expanding different types of irrigation in different parts of the country? What should be the strategy of future development and the priorities given to different sources and to better use of the existing facilities as against taking up new projects? What are the measures needed to ensure productive and yet sustainable use of water?

Utilizable Water Resources

The total surface run off in India was estimated by the Irrigation Commission of 1972 at 188.7 m. ha. m of this 66 m. ha. m was considered to be the ultimate potential for utilization. More recent estimates place the surface flow at roughly the same figure, but the estimates of potential has been revised slightly upward to 69 m. ha. m. In 1970, a little over half of this volume is estimated to have been harnessed for use (Chaturvedi, 1976). Assuming that the ratio of utilization to storage has remained constant, the total quantum used in 1990 is estimated at 55 m. ha. m (Reddy, 1992).

The figures of utilizable flow do not include possible inter basin tranfers. Earlier 'visions' of interlinking the Ganga and the Cauvery mooted some years ago have receded into the background. However, the National Perspective Plan for water, prepared in 1980, envisages a beginning being made with the peninsular rivers. Specifically, it envisages the interlinking of the Mahanadi–Godavari–Krishna–Pennar–Cauvery, and the diversion of surplus west flowing rivers eastward. It is estimated that by doing so all the four southern states can get substantial additional supplies, totalling 58 bn. m^3 or 5.8 m. ha. m., capable of irrigating an extra 9 to 10 m. ha.

The annual groundwater recharge for the country was assessed to 42

m. ha. m. in 1972, the most recent estimates place it at 45.2 m. ha. m. The utilizable volume which was reckoned at 26 m. ha. m. in 1972 (about 60 per cent of recharge) has now been revised to 38–9 m. ha. m (85 per cent of recharge); actual utilization as of 1990 being about 30 per cent of the revised potential.

In terms of gross area irrigated, the ultimate potential, according to estimates published in the Eighth Plan document, is of the order 114 m. ha. compared to 64–5 m. ha. reported to be actually irrigated in 1989–90. At the current level of irrigated cropping intensity, this would imply 88 m. ha. or nearly two-thirds of the country's cultivated land could get the benefit of irrigation. Before taking comfort from these figures, however, they require a closer critical look.

The estimates of available utilizable and utilized water resources leave much to be desired. The estimates of surface flows continue to be based largely on empirical formulae relating rainfall to surface run-off pioneered by A. N. Khosla in the 1950s. The lack of data based on measurement of actual flow in the main river and tributaries of different basins over sufficiently long periods—30–40 years observations is usually considered to be reasonable—remains 'one of the most serious handicaps in the planning of water resources development.' Though the Central Water Commission has established a hydraulic observation network all over the country, it is far short of standards specified by the World Meterological Organization (Reddy, 1992). The States have their own gauges, but, since many rivers are the subject of inter-State disputes, they are unwilling to provide the data on flow measurements. Systematic analysis of the deviation between yield asssumed at the time of project design and record of actual receipts in completed projects could provide the basis for improving the empirical basis for estimates of surface flow–rainfall relations. This has however not been done.

Estimates of utilizable surface flow are nebulous both conceptually and in relation to their empirical basis. One could in principle think of utilizability purely in technical terms. This would require detailed surveys to identify all the possible sites for storage or diversion of river flow, evaluating these sites from the viewpoint of engineering feasibility, and the quantum which could be drawn for use. Such an exercise was attempted in the CWC during the fifties, and reports indicating potential sites, their storage capacities, and utilization were prepared for major river basins and their sub-basins.[12] The reports have not been published, nor does the exercise appear to have been followed up by detailed field surveys and investigation to improve and refine the estimate.

What is technically feasible may not be economically viable if the costs involved are not justified by the likely returns. By the latter criterion, utilizable flow would generally be less than what is technically feasible. However, economic viability itself is a function of the state of technology and engineering of irrigation design and construction, technology of water management, as well as developments in agricultural and water application technology. The economically feasible or justifiable potential would therefore be changing over time. The CWC estimates of utilizable flows do not take these considerations into account. They are based primarily on technical assessments which, as mentioned above, are neither complete nor thorough.

The estimate of ultimate potential in terms of gross area irrigated is subject to further caveats. As has already been noted, the extent of area irrigated from a given amount of water depends on a complex of agro-climatic factors, the nature of the system, crop patterns, and standards of water management. The necessary information on all these planes are not available even in relation to existing projects. At any rate, the transformation of volumes of utilizable water into area irrigated does not even attempt to grapple with the complexities involved in the exercise. For all these reasons the estimates of 'ultimate potential' of surface water are at best indicative and should be treated with a great deal of circumspection. I have already detailed the limitations of the available estimates of the actual utilization of water and area irrigated from different sources and for different users.

Groundwater recharge consists of two distinct elements: recharge from rainfall and seepage of applied water in areas irrigated by surface sources. The replenishable recharge from rainfall depends on numerous factors: the level and intensity of rainfall, the density of vegetative cover, land slope, and depth of the soil. The magnitude of recharge can be increased by soil and *in situ* moisture conservation measures. These parameters and the extent to which they can be manipulated vary a great deal both between and within river basins. Our knowledge of these determinants is not sufficiently detailed and accurate to permit confident assessments of the extent of natural recharge.[13]

Recharge through seepage from surface irrigation, being a function of the total volume of water supplied through surface works, is somewhat easier to estimate. However, in the absence of systematic and precise measurements of seepage and its behaviour in different systems, the estimate can only be approximate. Nevertheless one can safely hazard the prediction that its magnitude will increase with the spread of surface irrigation.

Constraints on Development

The development of water resources is a State subject and, with few significant exceptions (e.g. Bhakra Nangal and DVC), it has been largely left to the initiative of the States. Since irrigation generally makes for a substantial rise in productivity, there is widespread interest among farmers in irrigation development. The government is under great pressure to construct more irrigation projects on its own and to facilitate private irrigation development (mostly through wells) in various ways. Each State government is keen to be seen as an active promoter of irrigation for the benefit of as many of its constituents as possible.

Most of the rivers in India, however, flow through more than one State, and the availability of water in relation to the cultivated area is both limited and highly variable. So is the extent to which various States have exploited their 'potential'. Since, even on full development, a substantial part of the cultivated area in all the States will remain rain-fed, and since some regions have, for historical reasons, lagged behind others, there is naturally a desire on the part of every State to appropriate as much water as it can and to develop irrigation as rapidly as possible.

In the process, disputes about the appropriate sharing of the river waters have come into sharp focus and intensified. Central government clearance is required in the case of major projects involving inter-State rivers. In a few cases (e.g. Bhakra Nangal, Narmada, and DVC), the Centre was able to bring riparian States together for the combined development and use of the water of major rivers. The functioning of these arrangements has not been free of problems. In 1956 a law providing for the constitution, by the Central Government, of River Boards for the regulation and development of inter-State rivers and river valleys. The Boards were to advise the concerned riparian States on coordinated development and optimum use of the basin resources, help resolve disputes, as well as undertake basic technical studies, prepare plans, and advise in their execution. The necessity for basin-wide planning has been often repeated since, the latest occasion being the National Water Policy statement of 1986. Very little has however been done to implement the idea. So far not a single river basin board empowered to take up integrated development of water has been set up.

Around the same time as the River Boards Act was passed, the Central government also enacted a law for adjudication of inter-State water disputes through tribunals. The Centre has chosen to rely on this instrument, though not in all cases and consistently, to tackle disputes rather than create a continuing organization of the type visualized by the

River Boards Act. In the case of the Krishna, Godavari and Narmada tribunals appointed under the inter-State Water Disputes Act have given 'awards' indicating the extent of water that may be utilized by each of the riparian States.[14] In the case of Cauvery, the pre-Independence developments in the lower part of the basin were governed by an agreement between the old Madras Presidency and the old Mysore State in the 1920s. This agreement is now due for a review. There is a serious dispute as to the amount of water Tamil Nadu is entitled to under the present circumstances. Negotiations which have been underway in fits and starts for several years have not produced any agreement,[15] and yet there is a remarkable reluctance on the part of the Centre to invoke the process of adjudication under the Inter-State River Dispute Act.

But, as the Irrigation Commission rightly pointed out, while tribunals and other judical/quasi-judical bodies deal adequately with the legal aspects, disputes relating the sharing the water of a river cannot be satisfactorily resolved by a judicial decision.

Inter-State river disputes are seldom so clear cut as to admit of unequivocal decisions. There are also a multitude of variable and imponderable factors which have to be taken into account, some of which already exist and some of which may arise in the future. For this reason, the apparently simple and justiciable issues of sharing of water may get involved in a complex of socio-economic issues of greater moment than the simple technical issues involved. [GOI, 1972.]

The tribunals have done the best they can with the information and expertise available to them. Both these are however far from adequate. At best their awards provide a starting point. It is extremely important to have a continuing organization to generate the basic data, monitor the evolving pattern of water use and their impact, and to periodically review the plans in the light of accumulating experience. River basin boards with the necessary autonomy, authority, and expertise are a precondition. At the same time, it is important to make the relevant information available in the public domain, and to encourage involvement of non-governmental experts in analysing the data and exploring the optimum patterns of development and use of water. Such analyses from different viewpoints make for better informed deliberation of the choices.

Inter-basin transfers and cooperative development of the Ganges–Brahmaputra water resources by the riparian states raise even more difficult problems. The Central government set up the National Water Development Agency to initiate a detailed study of the scope and feasibility of inter-basin transfers.[16] Studies on the Mahanadi–Godhavari–Krishna link and of diversion of some west flowing rivers were taken up in the Seventh Plan. States like Tamil Nadu and Karnataka are naturally are keen to have this study completed early, but others are less enthusiastic. There is a dispute regarding the existance of any surplus to spare; there is also an apparent conflict between using west flowing rivers to generate hydel power on the western side of the Ghats and diverting it to the east. It has been suggested that this need not be competitive, inasmuch as power can be generated from the water diverted to the east as well.

Cost seems to be a serious problem; the estimated cost of inter-basin transfers is Rs 80,000 per ha. which is much higher than the cost of providing irrigation by using existing supplies in Andhra Pradesh and Karnataka. The economic viability of such schemes, and the manner of financing them, are thus major issues. There are also questions concerning environmental aspects, especially on submergence of forests and displacement. The legal and institutional problems of enforcing inter-basin transfers across inter-State rivers are also far more complex than within particular inter-State river basins.

The inherent conflict of interests, doubts about the impact of inter-basin transfers on the availability of water to riparian States which have not yet exploited the feasible opportunities within their territories, and the high level of politicization of such issues are serious impediments to vigourous action on the part of the governments—be it the Centre or the States. It is highly desirable that non-governmental organizations and non-official experts should get together and apply their minds to an objective, professionally competent study of the problems of optimum development and use of water in the major river basins. They should put together all available information on the water resource potential, the present level of water use and its efficiency, and possible ways of harnessing a larger volume of water and to raise the efficiency of water use; and alternate strategies for exploiting the resources and their implications in terms of feasibility, costs, returns, and the distribution of benefits. These should be made public and discussed in public fora in order to promote an informed discussion of policy options. This would hopefully facilitate political compromise with a wider regional/zonal perspective than is likely under existing arrangements.

Prospects of Reducing Regional Disparities

The bulk of the unexploited potential, according to official estimates, happens to be in States with a low level of irrigation development and/or a low level of land productivity. The interests of reducing inter-regional disparities in both respects would therefore be served by measures to speed up the harnessing of the potential. This would however call for a significant change in the distribution of additional irrigation investments across States, and even within States. Irrigation is a State subject and most of the investments in this sector are components of the State plans which have to accommodate the claims of many other programmes. States with large unutilized potential are in many cases in a weak financial position.

Over the past three decades or more, the relative share of different States in total public sector outlay for irrigation and flood control has varied within remarkadly narrow limits. The central government has been providing extra assistance for speeding up completion of ongoing projects, but this amounts to a fraction of the investment under State plans and can make only a limited impact. Institutional finance for groundwater schemes of the private sector can be used to help financially weak States to access more resources for irrigation development, but insofar as increased availability of groundwater depends on the expansion of surface irrigation, the scale of public investment is crucial.

Finance apart, narrowing regional disparities in the level of irrigation cannot be the sole criterion for deciding the level, composition, and regional distribution of investments in the sector. It is necessary to consider the costs of irrigation in relation to benefits; the relative merits of improving the efficacy of existing works as against construction of new projects, and of investments in rain-fed as against irrigated agriculture. Roughly two-thirds of the irrigation potential is in States where the ratio of irrigated to unirrigated yields is less than 2, and only one-third is in States where the ratio exceeds 3. Again, about one-third of the unused potential is in States where *public investment* in irrigation per unit of addition to output from irrigation exceeds 10, and around 30 per cent in States where the ratio is less than 5 (Table 18). The eastern region, which has a large part of the unsued potential is marked by a high public investment to output ratio for irrigation. These comparisons are not conclusive, however, inasmuch as the estimated impact of irrigation on productivity relates to *all* sources and not to major and medium works alone. The estimates tend to overstate the productivity impact of major and medium works but not to a uniform degree in all States.

The economic justification for investments in such works will improve if their capital costs were reduced and they were managed more efficiently. There are, as has already been noted, strong reasons to believe that substantial cost reduction is possible if the mechanisms and procedures for planning, design, scrutiny, and award enforcement of construction contracts are made tighter and stricter. The necessary elements of these reforms are discussed later in some detail.

TABLE 18

Indicators of the Potential, Cost and Productivity of Irrigation, by State

	Avg. cost of additional potential (Rs per ha.)[a]			Productivity diff. between irr. & unirr. land*		Investment in irr. per unit of additional output	Irrigation potential (1,000 ha.)	
	Major	Minor	All	Absolute	Ratio		Created up to 1984–5	Ultimate
Andhra Pradesh	15,200	2,260	9,780	1.237	3	8	5,600	9,200
Assam	5.208	6.770	NA	NA	NA	NA	520	2,670
Bihar	12,690	1,660	6.030	600	2	10	6,300	12,400
Gujarat	37.690	4.630	23.730	1.250	3	19	2,940	4,750
Haryana	24.130	1.580	12.860	1,355	4	10	3,326	4,550
Jammu & Kashmir	15,225	1,480	15,050	NA	NA	NA	NA	NA
Karnataka	11.000	3,100	7,510	1,265	3	6	2,400	4,600
Kerala	12.190	4,000	9,550	NA	NA	NA	950	2,100
Madhya Pradesh	14,700	5.360	10,170	605	2	17	3,790	10,200
Maharashtra	20.700	4.990	14.810	1.324	4	11	3,700	7,300
Orissa	14,400	2,070	6.740	389	2	17	2,630	5,900
Punjab	16.290	400	7.000	1,465	3	5	5,640	6,550
Rajasthan	11.360	2,270	8,520	898	4	10	3,800	5,150
Tamil Nadu	21.400	2,320	7.880	1,677	4	5	3,190	3,900
Uttar Pradesh	8.750	840	3,750	983	2	4	18,800	25,700
West Bengal	12,000	3.060	5,650	NA	NA	NA	3,300	6.110

Note: NA—Not Available.

a—Public sector outlay during 1980–85 per hectare of increase in irrigation potential during the same period. Due to difference in size, date of start, and mix between completed and ongoing projects, the estimates may not give a correct picture of relative costs.

*—Value calculated at the avg. of all-India prices, in 1970–73 as estimated by Bhalla and Alagh (1979).

Ratio: per net irrigated area.

Absolute: Gross irrigated area.

Source: A. Vaidyanathan (1987); and Centre of Monitoring Indian Economy (1988).

PRIORITIES IN IRRIGATION DEVELOPMENT

The priorities to be accorded to improvement of existing systems as against construction of new works, and between large surface works in relation to minor irrigation, are also at issue. Modernization of existing systems is essential in most cases to improve control over water deliveries. The magnitude of investment involved is however relatively small in comparison to new projects. According higher priority to the former need not cut seriously into allocations for new projects. The more important, and difficult, prerequisite for improving water management lies in many cases in origanizational and managerial reform.

There is a view questioning the importance given to major and medium surface irrigation in public investment on grounds of economy as well as the environment. Small surface works, it is argued, are cheaper, take less time to complete, and are more conducive to management by user communities. Groundwater irrigation, by its very nature and because it is in the hands of individual farmers, is deemed to be more efficient. These propositions are however misleading and open to challenge on several grounds.

Thus, cost per hectare of potential created by minor surface works does not allow for works going out of use (because of neglect, conversion of tank beds into residential colonies, and such other factors). The reported investment in minor works is more in the nature of gross investment. If allowance were made for works going out of use, the cost per hectare of net addition to potential would be much higher. On the other hand, if the area irrigated by these sources has in fact declined (as LUS data report), the investments have not produced any net benefit at all!

In the case of groundwater, the full cost of development cannot be estimated from government expenditure on it or from loans by public financial institutions for the purpose. Apart from the need to allow for infructuous drilling and for existing wells going out of use, the substantial, but unquantified, volume of private investment in well irrigation (including construction of distribution networks) needs to be reckoned in the calculations. Moreover, well irrigation requires investment not only in wells/bores, pump-sets, and structures but also in transmission lines and power generation (in the case of electric pumps, and in production and refining of crude oil and distribution of diesel (in the case of oil pumps). The latter indirect investments are essential to sustain groundwater irrigation and should be counted in any meaningful measurement as part of the investment costs for this sector.

There are, however, other difficulties that vitiate meaningful comparisons of the relative merits of different categories of works on the basis of cost per unit area irrigated. This is because of differences in the seasonal patterns and reliability of supplies between sources, and also in the extent of conveyance and application losses. The supplies from tanks are more concentrated, less reliable, and usually available for shorter periods than that from large storages. Wells, on the other hand, are more reliable and conveyance/application losses are low. Obviously, when different sources are used in combination (e.g. system tanks and conjunctive use of groundwater in surface irrigated areas), comparisons of each type separately does not make much sense. Ideally we need to move towards estimation of costs per unit of water tapped and per unit of water effectively available for crops in different kinds of systems namely solely large surface works, small diversion/tanks, groundwater, and a combination of these. Differences in the reliability and seasonal distribution of supplies will also have to be taken into account while making comparisons.

In any case, the choice between different categories of works is not unconstrained. In any given tract, the effective choices between large and small storages used between surface and groundwater is limited by the local and seasonal pattern of rainfall, topography, and geology. It is no accident that small irrigation tanks are concentrated in those parts of India receiving rainfall from both the monsoons and where the topography is favourable for shallow storages. However, given the high seasonal concentration of rainfall in most parts of the country and the fact that rainfall tends to be higher in the upper catchments of river basins, monsoon rainfall has to be harnessed over a wide catchment and impounded on a large scale if it is to be effectively used for irrigation in the dry season. Under these conditions, large and small storages should complement rather than substitute each other. As for groundwater, the problem being one of checking over-exploitation, the need is to contain its use rather than expand it. There is, of course, scope for extending well irrigation in command areas of surface projects (especially new ones), but here the two sources are not alternatives but complementary to each other.

Having said this, there are compelling reasons for giving much greater attention and resources to small-scale surface irrigation schemes. The reported decline in area under this category of works is a reflection of past neglect. These works have not received much attention under the Plans, and investments in this category have been meagre in relation to be magnitude of the problem.

A major effort at *renewing and improving traditional local systems* is necessary. Such a programme should focus on the following essential elements: reforestation of the catchment areas of the tanks; restoring the inlet channels to their original capacity by clearing them of weed and silt and removing encroachments; strengthening and improving the tank *bund*s (or anicuts as the case may be), and other associated structures; and undertaking such improvements or corrections in the distribution network as the users feel to be necessary.

The user communities have first hand knowledge and experience of local conditions and local problems which no outside 'expert' can match. Their knowledge is valuable in deciding the improvements necessary in specific local contexts. Planned in consultation with user communities, 'modernization programmes' can be more effective and also much less expensive than the current approach which hardly involves the user community. The involvement of pre-existing local experience and institutions will also facilitate proper management of the 'revamped' system. Such a programme can be easily fitted into the employment guarantee and other schemes of land and water improvement as part of local area planning by *panchayat*s with technical help from irrigation departments and NGOs.

As for the *expansion* of minor surface irrigation, it would be wrong to assume that they are inherently and everywhere superior to large storage systems. Nor should they be viewed as alternatives. A more promising approach would be to explore the complementarities between small local storages and large reservoirs. The integrated operation of the two would mitigate the need for large storages and attendant problems; give greater room for meaningful user involvement in management; as well as greater flexibility and therefore more effective use of available water. The possibilities of linking up and integrating existing tanks with larger surface systems as well as planning new surface systems on the basis of integrated use of large and small storages therefore deserve to be exploited consciously to the greatest possible, degree.

One must also weigh the cost and benefits of irrigation expansion against those of rain-fed agriculture: while it is certainly true that investments in rain-fed agriculture do not appear to have made much of a difference to productivity, it is arguable that these investments have been badly planned and implemented, and that not enough is done to make them effective. The near stagnation of productivity of rain-fed agriculture, on which large segment of rural population depends, argues for much greater attention and resources to be devoted to this sector. Integrated watershed development, which seeks to increase *in situ* moisture conservation (by increasing the moisture trapped and retained in the soil or by devices like percolation ponds), is really another way of managing available water. It could open up significant opportunities for construction of new local irrigation works, besides improving the supply of existing areas.

The development of new small-scale water control works as part of the integrated watershed development programme holds vast promise. Integrated watershed development—which has to be planned at different levels from the small micro watershed covering a village or part of it to larger sub-basins and basins of river systems—provides an effective means of integrating afforestation, soil conservation, and percolation ponds and other moisture conservation works. Such integrated planning helps to make the investments for more effective, functionally and economically, than the present fragmented approaches to the problem. It is however also more demanding on both the government and the affected communities. They pose major challenges to institutional reforms, which are discussed in detail later in this volume.

ENVIRONMENTAL ASPECTS[18]

Concern over environmental aspects has been heightened by (a) the unexpectedly high, and in some cases alarming, rates at which reservoirs are silting up; (b) the direct effect that construction of large reservoirs has on the immediate environment (the submergence of economically and ecologically valuable forests as well as the displacement of people living in the area); (c) fear of over-exploiting groundwater; and (d) damaging cultivated land by indiscriminate irrigation. The environmentalists have marshalled a great deal of information on these aspects. Although the accuracy of the estimates may be open to question, and one has to guard against environmental fundamentalism, there is little doubt about the importance of the issues raised by environmentalists. The processes of environmental degradation are so prolonged and their consequences manifest themselves so slowly that they are dangerously easy to ignore or dismiss. The environmental aspect of water resource development and use must therefore receive much greater attention in the future.

It appears entirely sensible to insist that measures to check deforestation of upper catchments and to restore forest cover to an appropriate density must be an essential and integral part of planning for irrigation projects. Ideally, this should be done on a watershed basis, but this does not detract from the necessity of ensuring that the catchment of particular projects are protected. The principle is now recognized, but apparently not taken seriously enough in actual planning. One way of generating a greater sense of seriousness is to insist, as a precondition for considering all proposals for new irrigation works, that the project design state the conditions of the catchment and propose concrete programmes to remedy deficiencies.

The loss of forests and displacement of people is a different type of problem. There are those who accord such absolute priority to preserving forests and the life-styles of the people inhabiting them that they oppose all large reservoirs. They might concede that if irrigation is necessary, it is preferable to rely on such solutions as a number of small, local storages or groundwater development that do not damage the forests and do minimize human problems arising from displacement. Although much can be said for tapping water in an environmentally safe manner, it is not always feasible. For instance, as already noted, the Deccan has relatively little groundwater, and small storages that depend on rainfall in their respective catchments will not significantly augment the supply of water for crops in the non-monsoon months when it is most needed. Relatively large storages that impound the surplus flows during the short monsoon season are a technical imperative. One could, however, conceive of fewer large storages combined with and linked to smaller ones. This possibility needs to be assessed in each specific context.

Unless there are ways of raising output significantly on drylands, reservoirs and the attendant submergence are a cost society must pay in order to feed its growing population. It is perfectly legitimate to insist that proper rehabilitation of displaced persons must be provided whenever a project leads to such displacement. The environmentalists' apparently extreme position stems perhaps from their frustration and anger with the tendency to pay lip service to resettlement issues while doing little about them. Their frustration is understandable, but the answer does not lie in fundamentalism. They should focus instead on ensuring, by legal and political means, that this aspect is attended to seriously and adequately.

These, as well as the problems of waterlogging and salinity and depletion of groundwater which have been discussed earlier, are receiving increasing recognition, thanks, in large part, to the effort of environmental organizations. The public debate on the Narmada and the Morse review of the project have played a valuable part in this, but much remains to be done.

One essential requirement is that preventive/corrective action should be an integral part of the design and implementation of projects. The Eighth Plan rightly recommends that measures for catchment area treatment and rehabilitation of people be explicitly incorporated in the project proposal backed up by necessary financial allocations and implementation machinery. Indeed, this should be made a precondition for a project to even be considered for investment clearance. It is also necessary to ensure that:

the crop and water use patterns are planned with due care and after proper investigation so that waterlogging can be prevented. The appraisal process should also try and make sure that the project design contains the disturbance of existing population settlements and forest cover to the minimum feasible within the available alternatives.

'The process of decision-making on these projects also need to be made more open so that the public at large, and in particular those directly affected by a project, can have access to adequate information about the assumptions and calculations on which a project is judged by the authorities to be technically and economically viable; to satisfy themselves that sufficient safeguards have been built into the project to take reasonable care of those who are affected by the projects and also the potential adverse ecological consequences flowing from the construction of the project and its operation; and give them an opportunity to place their objections and concerns before the concerned authorities along with concrete suggestions for alternative, cheaper/safer ways of achieving the objectives which the project is supposed to serve' (GOI, PC, 1992).

If these admirable sentiments are to be made effective, it is essential to evolve a procedure by which the techno-economic evaluation of all projects with a displacement effect or environmental impact are required to be made available to people in the areas affected by the project to elicit their reactions. A reasonable period of time should be allowed for the people to comment on and raise questions regarding the project, and also offer suggestions on alternatives. The decision-making authorities should be required to give due consideration to these reactions before taking a final decision. It is important that the government agencies should be seen to be sensitive to the legitimate concerns of the affected region or people. It is equally important to evolve some mechanism for continued monitoring of the implementation of the project and for speedy redressal of grievances about non-compliance. The approach suggested here would make for more informed public discussion of issues, and also serve as a salutary check on any tendency to approve ill-prepared projects without due scrutiny.

NOTES

1. In the latter case, the Irrigation Department's data are likely to cover only public irrigation works. Data for irrigation under private works, and by works for which no water charges are levied, are presumably collected by the *patwari*. If the latter maintains a record of all the irrigated area, this in principle would act as in an independent check on the Irrigation Department figures, but it is doubtful if this cross check is in fact made systematically and objectively.

2. The National Sample Survey has in fact been giving decennial estimates of land use and irrigation. There are wide differences between LUS data on the relative importance of surface and groundwater and estimates from the National Sample Survey.

	1953–54		1981–82		1991–92	
	NSS	LUS	NSS	LUS	NSS	LUS
1. Net irrigated area as % of net sown area	14.8	17.3	29.7	29.0	35.4	32.3
2. Percentage of NIA under groundwater	32.9	29.0	55.6	45.6	54.4	59.1
3. Gross irrigated area as % of gross cropped area	14.7	17.1	29.7	29.7	–	33.3

It will be seen that the share of groundwater irrigated area according to LUS is consistently less than in the NSS. The latter shows a faster rise in the net and the gross irrigated ratios as well as the area served by groundwater in comparison to the land use data. This has not however attracted much attention from officials or researchers.

3. Data by basin on live storage capacity of reservoirs in the 1980s have been published (Reddy, 1992): The aggregate storage capacity of reservoirs in the country at the end of the 1980s is estimated at 193 bn. m^3 in comparison to 140 bn. m^3 in 1970–1 and around 30 bn. m^3 at the time of Independence. Total utilization in 1990 is estimated at 550 bn. cu. m. (i.e. 2.85 times the storage capacity) (ibid.). On this basis the gross utilization of surface sources in 1990 works out 350 bn. cu. m.

4. Dhawan's estimates are in terms of 'foodgrain energy equivalents' (based on the calory content of different crops). Vaidyanathan's estimate is based on gross value of crop production at farm harvest prices. For a more detailed discussion of alternative ways of estimating the impact of productivity, see Vaidyanathan et al. (1994).

5. For a good review of available studies see Pant (1982).

6. The CWC cost index has been extended to a more recent period in Rajagopal and Vaidyanathan (1997). It also discusses the limitation of the Index. The Ninth Plan working group on irrigation also provides estimates of outlays at current and contant prices, up to the early 1990s. See also Gulati and Svendsen (eds) (1994).

7. Note that except in Tamil Nadu, official sources do not distinguish between area irrigated by groundwater as a sole source and in conjunction with surface sources. Nor do they clearly spell out the basis of classification by source so that possible double counting is avoided. This limitation of the data on irrigation by source suggests that the data need to be interpreted with caution.

8. See Vaidyanathan and Janakarajan (1986), Sivasubramanian (1998), Vaidyanathan (ed.) (1998). These are all as yet unpublished studies done at the MIDS under sponsorship of the Planning Commission and the Ministry of Environment and Forests.

9. See GOAP (1982), Wade (1980), also Vaidyanathan and Janakarajan (1986), Rajagopal (1996) for a description of a few large surface systems. Also see Ch. 1 of this volume.

10. This section draws heavily on the Report of the Committee on Irrigation Pricing, GOI, PC (1992).

11. The figures cited here (and Table 16) are estimates of the National Institute of Public Finance and Policy (Mundle and Rao (1997) and Srivatsava and Sen (1997)). They substantially differ from the estimates of the Committee on Irrigation Pricing (GOI, PC, 1992). The latter came to a figure of Rs 15,255 m. as unrecovered cost on account of irrigation in 1986–7, while the Mundle–Rao estimate for 1987–8 is Rs 47,000 m. The difference is principally on account of the fact tht the Committee on Irrigation Pricing used the cumulated capital outlay on major and medium works three years prior to the year for which the subsidy was estimated. The NIPFP, on the other hand, used the cumulative government outlay on irrigation up to and including the year for which estimates were made.

12. These reports, running into several volumes, give assessments of technological possibilities of irrigation development for all river basins and sub-basins. They include maps indicating locations of existing and potential storage sites. The reports, which were put out, are not readily available. They have not been updated.

13. The deficiencies in groundwater potential estimate were explicitly noted in the Eight Plan document (GOI, PC, 1992).

See Dhawan (1996), for a detailed critique of groundwater potential estimates and their rationale. The methodology of estimation has been reviewed recently by a Groundwater Board expert committee.

14. See Ramaswamy Iyer (1994) and Chauhan (1992) for a review of dispute settlement procedures and their working.

15. See Guhan (1994) for a detailed account of the historical background and current status of the Cauvery dispute and the issues involved.

16. For current state of work on this aspect see Anon (1996).

17. The treatment of environmental aspects of water resource development in this paper does no more than highlight the essential problems. The treatment is rather general because of the paucity of reasonably comprehensive and reliable information on practically every aspect—be it the quantum of resource, patterns of utilization, costs of exploitation, the pace of degradation, its attendant consequences and costs of correcting them. The various environment groups have brought together a great deal of available information, but the fact remains that

the infomation is considerably less than what is needed for informed intervention. Also in a country of India's diversity any meaningful consideration of the environmental dimension must necessarily deal with it at a region specific level. A major effort to identify the key data required for environmental monitoring and organizing it as a continuing, institutionalized activity is essential. Much can be done through the NGO networks both to collect the data and to devise cost effective methods of getting the data through trained local observation groups. The latter is potentially a very effective way of raising public awareness on a wide scale, making people think about the environmental problems, and the necessity for farsighted interventions as well as the critical importance of active but informed local involvement in any programme of remedial action.

For detailed discussion see Vaidyanathan (ed.) (1998).

3

ASPECTS OF INSTITUTIONAL REFORM IN INDIAN IRRIGATION

INTRODUCTION

E ven as the achievements of the past five decades in irrigation development are considerable, there are clearly many lacunae and weaknesses in planning, implementation, and management of water resource development. Planning is fragmented on the basis of size (major and medium projects as against minor works) and sources (surface and groundwater). For the most part, water resource development is planned on a project-by-project basis, focusing on particular uses. Requirements for non-irrigation uses are paid insufficient attention in project design. Engineering aspects tend to dominate project planning over determination of optimum crop patterns, irrigation schedules, and other aspects affecting the effectiveness of irrigation. The mechanisms for scrutiny and appraisal of project proposals, and monitoring of construction are quite lax. System facilities are not maintained in good repair. Irrigation schedules are unpredictable, and complaints of inadequate and uneven distribution of water widespread. The productivity of irrigated agriculture is much less than the potential. And the high degree of politicization of managerial decisions and their implementation has resulted in higher-than-necessary costs, lower-than-feasible productivity, and very low cost recovery.

Investment in technical improvements (such as realignment and desilting/lining of canals, more and better regulatory structures, better crop planning and field irrigation techniques) are of course essential, but not sufficient to correct many of these deficiencies. The way the various functions involved in irrigation are organized and managed are just as important. Indeed, the prospects of many of the desirable technical changes themselves hinge upon restructuring of institutions. This chapter deals with the institutional changes needed to improve (a) planning and management of surface irrigation; (b) *in situ* moisture conservation through integrated watershed development; and (d) regulation of groundwater in the interests of sustainability.

SURFACE IRRIGATION[2]

Investment Planning

The responsibility for planning and construction of surface irrigation projects vests almost exclusively with the Public Works or the Irrigation departments of State governments. Most of them have special divisions or directorates to perform these tasks. The Central Water Commission provides expert advice in matters relating to project design. It is also expected to review all projects on inter-State rivers and to appraise the technical and economic viability of major projects through its Technical Advisory Committee (TAC). Cost revisions of over 10 per cent are subject to its review. Clearance of the TAC is essential before the Planning Commission approves a project for inclusion in the Plan. The process however does not work satisfactorily.

Scrutiny is perfunctory partly because of inadequacy of data and poor institutional memory of past experiences and their lessons. The principal reason is however the laxity in mechanisms and procedures for evaluation of project proposals. The entire process is strictly internal to the government agencies. The criteria used are ill-defined and their application far from strict or rigorous. There is no mechanism for public scrutiny, nor are the details of the appraisal available to interested citizens and organizations to permit relevant issues being raised in the media and other public fora. Also, political pressures exert far too great an influence on decisions.

Correcting these deficiencies obviously requires a considerable strengthening of the technical organization for survey/design, in terms of expertise both in engineering and other disciplines (especially soil science, agronomy, environmental science, and the social sciences) relevant for proper planning of water resource projects. The scope of project planning *needs to be enlarged to cover uses other than irrigation and to ensure that the effective use of water is given as much importance in drawing up projects as the purely engineering aspects.* The aim of planning must be to match the different uses of water with different sources of supplying it optimally not only in the command area of particular projects but across all projects in a river basin. The appraisal process needs to be both stricter and more open.

Water resources available in any particular project location are derived from rainfall, stream flow, and groundwater, and there are many claimants (agriculture, domestic consumption, power, and industries) for its use. Depending upon the agro-climatic conditions and the nature of the terrain, the potential supply of water from the various sources, their relative importance, the technical feasibility of harnessing them at a particular site, as well as the needs for different uses in the command area, vary. The nature and magnitude of sources of supply, the techniques of harnessing them, and its allocation between different uses are all interrelated. The costs and benefits also vary, depending on how these elements are combined. Planning for the efficient and, at the same time, socially preferred development of available water—be it in a particular project or in a river basin—must therefore transcend narrow engineering perspectives and adopt a comprehensive systems view.

Experience with Command Area Development and the National Water Management Project has helped irrigation engineers to become more conscious of the need for a broader perspective in the planning and management of irrigation. The National Water Policy statement of 1987 emphasizes the need to recognize that water is a scarce and precious national resource, and that the needs for uses other than irrigation (and especially drinking water) should be given due weight in project design. The Planning Commission now requires preparatory studies of much wider scope covering not only hydrology and civil engineering aspects, but also of soils, land capability, and the drainage and groundwater conditions in the command area of the proposed projects. This has been done in some large-scale projects (like Sardar Sarovar and the Indira Gandhi Nahar projects) largely with the help of consultants from outside the Irrigation department.

The importance of integrated basin-wide planning of water resource development is also recognized by the National Water Policy and is reiterated by the Eighth Plan document. Some limited progress has been made due to the efforts of the Central Water Commission and the National Water Development Agency. River Boards/Commissions set up under the auspices of CWC have prepared studies on the development of the Ganges, the Brahmaputra–Barak, and the Son river basins. The NWDA has been studying the possibilities of inter-basin transfers of water and has prepared some studies as a basis for discussion. During the eighties a number of State governments are also reported to have prepared master plans for development of water resources in their river basins. (GOI, PC, 1992: 66.)

These initiatives have however only had limited impact. Basin boards have been set up only for a few rivers, and even these are ad hoc bodies. They have scarcely any role or responsibility for actual planning, and the various studies prepared by them do not enter in any significant way in determining the strategy, priorities, or project selection of the concerned States. Also since none of the above-mentioned basin studies have been published, or are even accessible to interested persons/organizations outside the government, it is impossible to judge their quality. The Planning Commission's guidelines regarding preparatory studies are not taken seriously and implemented by the State governments in project planning. Nor are Planning Commission and the CWC sufficiently strict in refusing to entertain proposals that are not supported by the prescribed preparatory studies of acceptable standard. There is no indication that survey and design organizations in the States have been expanded and made broad-based. States remain reluctant to accept the idea of integrated basin planning, not to speak of setting up organizations with the resources and authority to operationalize the concept, even in relation to basins that fall entirely in their domain.

There are understandable reasons why water resource planning with broad scope and with a long term perspective is so slow to take root. Part of the explanation is fragmentation of responsibility for irrigation, urban water supply, and meeting the needs of industry. Planning is largely in the hands of engineers, and the government engineering establishment is rather closed to ideas. Survey and design are not seen as a professionally challenging and creative activity deserving special encouragement. Interest in developing broad-based professional competence necessary for better planning is conspicuously absent. The present modes of organization and functioning of the PWDs do not give much encouragement or scope for inducting up-to-date techniques and specialized skills from outside.

Moreover, integrated planning for harnessing water from all sources (namely rainfall, stream flow, and groundwater), and to ensure optimal allocation between different uses and users, is a technically demanding task. Techniques for system modelling and algorithms for their estimation are of course available and advancing rapidly.[2] Their use however requires specialized training and a concerted effort to draw upon knowledge and expertise from several disciplines other than engineering. The task is difficult enough in the case of individual projects. Clearly when the plan is to cover an entire basin, instead of a particular storage site, the range of alternative use patterns and methods of harnessing and distributing water and combinations of these is much wider. Consequently, the scale and complexity of the exercise is also vastly greater.

The application of improved techniques of water resource planning is constrained by data availability.[3] The more sophisticated the techniques, the more demanding are data requirements. Even after five decades of 'planning', the information on which projects are designed remains incomplete, inadequate, and unreliable. Not that efforts have not been made to improve the database. The network of meteorological observatories and gauging stations to measure river flow has expanded. The Command Area Programmes were expected to generate more and better data on conveyance and application losses in the surface irrigation commands. The exploratory activities of Central and State groundwater organizations were meant to provide a better basis for assessing extraction and recharge rates and other aquifer characteristics in different environments. The All India coordinated programme for research on water use, launched by the ICAR in the early 1970s, was designed for a better understanding of the relations between soil moisture regimes, irrigation practices, input use, and crop yields. These have not however yielded the expected results. Insufficient funding is one reason; but a far more important one is that the concerned agencies did not take these activities seriously. Data collection is lax; and such data as are gathered, are not systematically collated and analysed; nor are the findings fed into the planning process.

A substantially larger, and better organized, effort to broaden and deepen the database for water resource development is essential. The scale of the programme needs to be considerably enlarged. A substantial part of the effort must of course be devoted to extending and improving the network of meteorological observatories and river gauging stations, and basic studies of regeneration flows, groundwater recharge, and crop–water relations. The tasks of organizations responsible for the activities need to be clearly defined and provided the staff and funds necessary not only for gathering relevant data but also for their continuing analysis. The necessary resources are modest in relation to the overall magnitude of investments in the sector, and the potential gains in terms of better designs and savings in capital costs.

These programmes will necessarily take time before they produce significant results, nor will they provide all the data relevant for project planning. It is essential to supplement these with scientific studies of the experience of existing systems. Functioning systems can provide much useful information on many aspects relevant for project planning. Thus, the design of new systems can greatly benefit from documentation and analysis of the reasons for divergence between the assumptions in relation to water yield, water use, irrigation efficiency, crop patterns, and productivity underlying the original design of projects and the actual experience in the course of their operation. The Committee on Plan Projects (COPP), set up by the Planning Commission in the fifties, published first rate studies on the functioning irrigation projects, focusing on the deviations from design, their causes and consequences. But COPP has long since been wound up and nothing has been put in its place.

There is a strong case for reviving the practice—which ironically was discontinued with the advent of planning—of preparing 'completion reports' after a project is commissioned. Objectively examining the deviations from the original specifications, time schedules, and cost estimates of the project, and analysing the factors responsible for the deviations is valuable in improving project planning and avoiding past mistakes. A periodic critical assessment of the criteria and standards used in completed projects in the light of more recent developments in materials, technology, and construction techniques would also help in better, more economic project design.

Arrangements for study and analysis of the impact of irrigation on agriculture also need to be strengthened and institutionalized. During the 1950s and 1960s the Planning Commission sponsored a number of research studies to assess the impact of irrigation. But for the most part these were in the nature of baseline surveys; few repeat surveys have been done to assess the long term impact on productivity and the environment.[4] There has been some continuing activity in research institutes across the country (for reference see Ch. 2); but they are not designed to assess the way water is managed or to estimate the impact on production in different types of projects over a period of time. Their quality is also extremely variable. In any case, they do not seem to feed into, or have much of an influence on, the governmental machinery for the planning and administration of irrigation projects. A bigger, better planned, and continuing effort to comprehensively document the actual experience (in terms of the spread of irrigation, the way water is actually allocated and managed, as well as the impact on productivity and income distribution) in relation to the original expectations/assumptions is essential. Government departments are ill-equipped for this task. Collection and analysis of the relevant data by independent agencies (such as universities and research institutions) is conducive to greater objectivity and therefore deserves larger and more sustained support from government.

Significant progress is these directions is unlikely unless the organization for water resources planning is strengthened and made tighter. The present system in which projects are designed, scrutinized, and appraised by government engineers, must yield to a more broad-based and open system. The planning organizations also need to take a proactive role in promoting data improvement. The various specialized research institutions (both technical and socio-economic) need to be more closely involved in collating and analysing the basic data needed for planning specific projects as well as carrying out preparatory analyses relevant for good design. There is need for much greater interaction between design engineers in the government and those in the university system, including the engineering and agricultural universities. More active involvement of the engineering faculties in universities and research institutes in project design and in evaluation of designs, along with arrangements for a two-way flow of personnel between the government and the universities, will help enrich both the quality of university education and the design of projects.

A major effort is necessary to make practicing and potential planners of water resource development appreciate the need for the comprehensive systems approach to water resource development and to train them in the use of the techniques of systems analysis and modelling relevant for this purpose. The content of basic training in water resource planning offered in the universities needs modification: engineering students have to be taught to view water resources project design not just as a civil engineering problem but as one where the civil engineer has to be sensitive to several other facets requiring knowledge and expertise of other disciplines (such as hydrology, agronomy, ecology, and construction techniques) to be incorporated in the process. It is also necessary to introduce students to the more recent developments in related areas such as materials technology, remote sensing, and instrumentation; techniques of social cost benefit and systems analysis; and organization and management of water control systems. Few universities now offer such courses even at the postgraduate level.

Irrigation engineers already on the job need to interact more with agricultural specialists and social scientists. There is also a need for specialized courses to sensitise them to the need for a broader systems view of design and management, and to familiarize them with the relevant techniques. The experience of WALMIs (Water and Land Management Institutes) which were meant for this purpose, needs, to be evaluated so that they it can be reorganized and reoriented to achieve their original aims. At the same time, the resources and experience available in the engineering universities, IITs, IIMs, Water Technology Centres, the Agricultural universities, and other research institutes should be exploited to build pool of indigenous experts familiar with the latest developments. This pool can be drawn upon for advice on the design of specific projects, conducting applied research, and coordinating specialized in-service training for irrigation personnel.

Water resource planning has also to be much more open and sensitive to the social and the environmental impact of the projects. This is essential because of the legitimate and growing concern with the problems of displacement of people, the impact of degradation in the catchment areas on the life of the reservoirs, the relative merits of large versus small projects, and the potential damage to soil on account of poor irrigation management and the lack of drainage. Moreover, different ways of utilizing water and managing its distribution have different impacts in terms of both overall productivity and distribution of additional benefits.

It is therefore necessary to explicitly examine alternative designs and their consequences in terms of displacement of human settlements, potential environmental damage, as well as impact on productivity and its distribution. Furthermore, it is necessary to provide, explicitly and as an integral part of the project, reasonable safeguards to protect the interests of the displaced and of the environment (including adequate financial provisions) for this purpose along with credible arrangements to ensure that these conditions are fulfilled. The present practice of evaluating the viability of a project proposal on the basis of a single design and keeping the whole process strictly within the confines of government is incompatible with these requirements.

The effectiveness of these measures will be greatly enhanced if steps were taken to make the process more transparent and open to public scrutiny. Transparency requires that, (a) the information concerning the design of each project, its scope, the proposed water use and related regulations as well as costs are accessible to all interested individuals and organizations; (b) a credible mechanism (such as of public hearings) through which affected parties can seek clarifications, raise objections, ask for consideration of alternatives, and get an opportunity to be heard; and (c) the project planners be required to meet these objections and clarify doubts before a project proposal is considered for appraisal by the Planning Commission or financial institutions.

All these measures are entirely within the powers of the government, but the prospect of its taking the initiative to introduce them are low. Governmental agencies are not used to having their decisions examined and reviewed in public. Legislatures and the Comptroller and Auditor General, who are supposed to keep a watch over government decisions, have proved largely ineffective. Apart from apathy, the limited time and expertise available with these agencies for effective review is a severe constraint. Besides, governments have got away with not giving the information (or with giving incomplete, and even misleading, information), relevant for a meaningful scrutiny. This habit, born out of a defensive attitude, has hardened over time as the incidence of wilfully faulty decisions and negligence in implementation of decisions has increased. It will take a great deal of concerted pressure from the institutions of civil society and lending agencies to overcome these obstacles.

One would expect opposition political parties to be interested in raising these issues and exerting pressure for change, but this is not happening. Instead they tend to exploit particular scandals as a means of embarrassing the party in power rather than pressing for effective, institutionalized mechanisms for ensuring transparency in the formulation and implemention of all major government decisions. The NGOs, especially those interested in protecting the environment and promoting people's participation in development, have made some dent. Agitations over mining and deforestation in the Himalayas, and projects such as Silent Valley, Tehri, and Narmada, have succeeded in pressurizing government to commission independent reviews.

Recognition that such reviews should be institutionalized is now accepted at least by the Planning Commission. The Eighth Plan document states:

The process of decision-making on the projects need to be made open so that public at large and in particular those directly affected by a project can have access to more information about the assumptions and calculations on which a project is judged by the authorities to be technically and economically viable, satisfy themselves that sufficient safeguards have been built into the project to take reasonable care of those who are affected by the projects and also the potential adverse ecological consequences flowing from the construction of the project and its operation; and gives them an opportunity to place their objections before the concerned authorities along with concrete suggestions for alternative designs/ safer ways of achieving the objectives which the project is supposed to serve. [GOI, PC, 1992.]

The passage of 73rd and 74th constitutional amendments has also opened up opportunities for local communities to influence water resource project planning.

Though these are welcome developments, they are no more than a beginning. The government's approach is still ad hoc, hesitant, and altogether defensive. Legislations are incomplete and inadequate; their implementation has to overcome severe resistance from politicians, bureaucracy, construction contractors, and others who have a vested interest in continuing the status quo. In any case, legislation and enquiry committees are only enabling devices. They cannot be effective unless the government agencies are willing to make the relevant information freely available and the affected individual/communities, and the NGOs advocating their cause have the necessary professional expertise. The experience of the Narmada agitation shows that the present situation is unsatisfactory on both accounts. Even as public pressure for freedom of information has to be sustained successfully, professional expertise is necessary to examine complex technical aspects of projects and effectively contest the actions of government agencies in courts and public fora with reasoned critiques offering viable alternatives. The beginning of such mobilization are in evidence, but a great deal remains to be done.[5]

Recent controversies relating to major water resource projects (e.g. Narmada, Tehri) also highlight the fact that those contesting a particular project or class of projects focus principally on the adverse ecological consequences and on large-scale displacement and disruption of communities from their traditional habitats. The potential beneficiaries have been mobilized to contest these claims but rarely on issues relating to fair sharing of the benefits or to ensuring that the project is constructed at the lowest cost and managed efficiently once it is commissioned. Since the entire cost of resource development and operation is borne by the government, the potential beneficiaries are more interested in pressurizing government to provide cheap water rather than on ensuring a fair balancing of competing interests (including those who stand to lose from a project) and providing economical and efficient service. If users had a direct financial stake in the system, they would be more interested in ensuring that the projects are designed well, constructed speedily and economically, and provide efficient service.

MANAGEMENT OF SURFACE SYSTEMS

Deficiencies in design and construction naturally affect the quality of irrigation services in terms of the predictability, adequacy, and timeliness of water supplies to all segments of the command. The physical structures of both traditional and modern systems in India are in a poor state of repair because of long neglect of routine repair and upkeep. The distribution system and regulatory structures need considerable modification and improvement in the light of changes in crop pattern, irrigation practices, and, in several areas, the spread of conjoint groundwater irrigation. Substantial investments in system improvement are necessary for improving the quality of surface irrigation, and this must be given priority over the construction of new systems. This is not however sufficient unless accompanied by improvement in the organization and management of the continuing operational functions—namely maintenance, regulation of water deliveries, resolution of disputes, and adaption of practices in the light of changing conditions.

In this context, it is useful to distinguish between small, local systems (such as tanks) serving one, or at best a few, villages and the large canal irrigation works serving extensive areas. The former are generally relatively old, with a long tradition of management by the user communities. Though the state has assumed responsibility for repair and maintenance and improvement of the main structures, it is not involved at all, or involved only minimally, in water management within the command area. The latter task (including maintenance of field channels in the *ayacut*) is handled by informal community institutions on the basis of well recognized conventions. These institutions continue to function despite major changes in the size and composition of the user communities, techniques of cultivation, and conditions of water supply. This does not however mean that they have always adapted fully to these changing conditions, or that their functioning is uniformly effective.[6]

Reforms must be designed to revive and strengthen them so that they can function more effectively and flexibly. However, any attempt to force these community arrangements into a rigid, formal legal framework is undesirable. The state's role should be to facilitate and support their functioning with active help from NGOs. Where the state needs to play a much stronger role is in clearly defining the relative rights and obligations of user communities which are part of a larger integrated system as well as the rights of user communities with those outside; the basis on which these entitlements can be changed; and putting in place an institutional mechanism for enforcing these principles. In the past the state has not only neglected this function, but one can cite instances where it has flouted conventions and even judicial decisions.

Major and Medium Projects

As in the case of local irrigation works, substantial investments in repairs, realignment of distribution of networks, more sophisticated regulating structures, drainage facilities, and land improvement are essential. Considering that they now cover 29 m. ha. (nearly 40 per cent of the total irrigated area), and that better use of available facilities is probably more cost effective than creating new systems, improvement of existing systems deserves higher priority than the construction of new projects. Certainly far more resources need to be allocated for this purpose than in the past.

This will however not solve problems arising from fuzzy rules, weak enforcement, pervasive interference in the allocation and pricing of water, and lack of accountability on the part of the irrigation bureaucracy.

Several approaches to organizational reform are under active debate and experimentation. These include: (1) self-management by user communities; (2) conferring on users well-defined property rights over water supply from a system with freedom to buy and sell these rights; (3) techno-managerial reforms; and (4) varying combinations of the above.

The rationale for self management is that users who stand to benefit from a common source of irrigation have the incentive to get together to develop the source; agree on the basis for sharing the water as well as the costs involved in harnessing and distributing it; and work out a system of mutual monitoring of rule compliance and dispute settlement. The requirements of transparency, equitable sharing and accountability for rule compliance can be met under such an arrangement. However, the conditions necessary for effective collective action, which ensures equitable sharing of costs and benefits and yet avoids the 'free rider' problem, are not easily fulfilled even in small, relatively compact, and homogenous communities. Where communities are heterogenous and stratified, the difficulties are great. Also, when several communities are served by a common source, as is invariably the case in major and medium projects, more complex arrangements are essential to define the relative entitlement of different segments of users and ensure that operation rules are consistent with these entitlement as well as to devise a credible rule enforcement and dispute settlement mechanisms. Such organizations necessarily have to be more formal and have to depend on a bureaucracy.[7]

The distinctive character of major and medium surface systems in India is that they are in the hands of a bureaucracy entirely under the control of the government. They have little autonomy in formulating rules or enforcing them strictly. Water charges are determined and collected by the government, and the management is not accountable to users. Changes in entitlement as well as actual appropriation of water use are accomplished through petitions to the bureaucracy and, more commonly, through corruption and political influence. So far, attempts to reform the management of these systems have largely focused on upgrading and broadening the skills of system personnel, improving internal communication, introducing rotational water supply and, in recent years, by setting up water users' associations at the outlet levels. The deeper structural issues of autonomy and accountability have remained untouched.

There is now a strong convergence of opinion, among those who have studied the problem in depth, that managerial and financial autonomy are essential preconditions to any meaningful reform.[8] Ideally, the management of each system must be permitted to work out the allocation rules and schedules to ensure optimum use of available water subject to such broad guidelines as to social parameters (e.g. priority between uses, and consideration of wider and equitable sharing of benefits, sustainability, etc.) specified by the state. The system management should be empowered to apply the rules with scrupulous fairness without any interference from the government, political parties, or other external forces. The state must, on its part, support measures that management may have to take to ensure compliance with rules and to penalize violations. Financial autonomy implies that the management be required to generate sufficient revenue to cover its costs (including capital charges) and be given the necessary powers to set and collect water charges from users.

There is a school of opinion that the monitoring and regulation of water allocation can be tackled more efficiently, and without having to administer a rationing system, by making water rights marketable: i.e. by granting a specified entitlement in water to each user and permitting them to buy and sell these rights.[9] Tradeable water rights have been tried, and continue to be in the vogue in several parts of the US. Recently, a number of Latin American countries have launched reforms creating tradeable rights in surface systems along with insistence on a viable rate of return on investment as an essential condition for financing by the state or financial institutions and requiring potential beneficiaries to contribute a certain minimum fraction of the investment costs.[10]

A closer study would however show that the definition of water rights (involving specification of location, volume, timing, and duration) in irrigation projects is inherently difficult. Water rights of an individual farmer served by a canal system will not be meaningful unless they are derived from a well-defined, internally consistent set of allocation rules for the system as a whole. Moreover, these rights cannot be defined rigidly: given the uncertainties in the supply of water from the common source, flexibility to adapt to variable conditions of supply and of demand is necessary. The greater the need for flexibility, the more difficult it is to define, monitor, and enforce individuals' rights.

In a large system, technical problems make it impossible for rights to be effectively traded between all its segments. The problems get compounded when the number of users is large. For all these reasons, even if system managers have the autonomy and the capability to define individuals' rights and observe them, in reality the scope for trading (both spatially and in terms of volume) is likely to be quite limited. The 'market' solution is therefore unfeasible; at best its scope is likely to be very limited. Quantitative administered rationing of water among users is therefore inescapable.

The effective operation of a rationing system requires of course a clear and internally consistent set of allocation rules; a competent bureaucracy with the freedom to apply the rules strictly; and a credible commitment on the part of the state to support rule enforcement. There is clearly much need and room for improvement in all three respects. Attempts through the Command Area Development and the National Water Management Projects (NWMP) to tackle them have not been successful. The induction of agricultural and extension specialists under CADA has not made much of a difference to the quality of management. The NWMP has skirted the problem of formulating operational and allocation rules for optimum use of water and its equitable sharing among the *ayacutdars*. Autonomy remains a very far cry, and interference in water allocation continues unabated.

Experience has, at the same time, shown that it is unrealistic to expect the bureaucracy to monitor and enforce rules right down to the individual plots. This is both impractical and costly. Therefore a purely techno–managerial approach to institutional reform, focusing on engineering improvements and increasing the professional competence and autonomy of managers, is inadequate. Active participation of users in making and enforcing rules is essential to ensure wider acceptability of the rules among users, increase the effectiveness of monitoring rule compliance, and internalize incentives to reduce the costs of maintenance and operation.

Encouragement of user participation is now part of official Indian policy: the National Water Policy statement of the Government of India that declared 'efforts should be made to involve farmers progressively in various aspects of management of irrigation systems, particularly in water distribution and collection of water rates.' Over the past two decades a number of initiatives to set up and foster user associations have been taken up by the governmental and non-governmental organizations. Several States have also enacted legislation to promote user involvement in the management of state systems (for details see GOI, PC, 1992, Ch. 6, and annexes).

So far however these efforts have been limited to the formation of users groups at the tertiary level. They have been patchy and not conspicuously successful:

The area covered by these initiatives is very small, less than 1 per cent of area irrigated at present.... For the most part the outlet and canal committees are there only in name; their functions are vague; they seldom meet; they are not consulted on substantive issues; nor are department officers required to follow their advice. There is also considerable reluctance, if not opposition, from the operational staff of irrigation departments to involving users in management; and even users themselves tend to be apathetic to the idea. [GOI, PC, 1992: 126–7.]

User associations at the tertiary level cannot by themselves accomplish much in terms of improving the quality of irrigation. In areas depending wholly on direct canal irrigation, the outlet committees have little control over schedules and the volume of water deliveries. Except where canal supplies are impounded in local storages, the outlet committees have no flexibility in deciding when and how the available water is to be used. Not surprisingly, user communities are better organized and play a more active role in water management under 'system tanks' than in areas served directly by canals (For evidence on this see Vaidyanathen (ed.), 1998). But most system tanks predate the arrival of canal irrigation. The existing regulations do not permit users or user groups to construct new local storage or even to regulate canal supplies among their members. Moreover, so long as the government assumes the entire responsibility for maintenance, and canal water is provided practically free, there is little incentive for such associations to form. Such incentives for local collective action as may exist are weakened when deficiencies in canal supplies can be made up by tapping groundwater on an individual basis.

In sum, the success of active user involvement in water management at the outlet level depends to an important degree on, (a) credible assurance regarding the quantum and duration of water supply that each group can expect; (b) giving users groups the freedom to create local storages below the outlet and to determine allocations among their members; (c) insisting that these groups assume responsibility for maintenance and ensuring that users pay for the cost of providing them; (d) vesting the powers for regulating conjoint use of groundwater in the outlet command to the respective user groups; and (e) direct participation of user groups and their representatives in the management of the system with autonomy to decide, monitor, and enforce operational rules at all levels of the system, to levy and collect water rates and apply them for better operation of their system (see Patil and Lele, 1995).

In such an arrangement, the state might finance part or all the costs of development, but on condition that (a) the system reimburse part or all of its contribution in a specified manner; and (b) the state will no longer determine the rules or interfere with their implementation. System rules are left to be decided by negotiation among the principal stakeholders (useres, managers, and the government) in the system. The opportunity to take a direct and active part in the process also strengthens the incentives for monitoring rule compliance, maintenance, collection of water charges and minimization of the costs of upkeep and operations. The state will continue to have a key role, but of a different kind: it will set up independent bodies to adjudicate issues concerning the relative claims of different uses and users both within the across systems, in a basin; regulatory authority to ensure reasonable and fair pricing of water; and provide strong support and the assistance of state power in enforcing the decisions of these bodies as well as the management of the irrigation systems.

Based on the above reasoning, an expert committee appointed by the Planning Commission to review Irrigation Pricing Policy, recommended major changes in the management of public irrigation systems (GOI, PC, 1992). It was of the view that a mechanical approach to raising water rates to cover costs is undesirable, and that these rate adjustments should be accompanied by credible measures to improve the quality of irrigation service, reduce costs, and ensure accountability. The principal elements of the reform suggested by the Committee include:

(a) Constitution of the management of each system into an autonomous authority with the freedom to decide its internal allocation and pricing rules subject only to broad principles and review by an independent regulatory authority.

(b) While the government may continue to provide investment funds, the users will also be required to finance a part of the costs. The management would be required to reimburse all or a part of the government contribution over a period of time to be specified by the government.

(c) The government, by legislation, needs to appoint adjudicatory and regulatory authorities: the former to decide on disputes over rights of access and use, the latter to review and approve management decisions regarding water rates charged by system.

(d) While the government, as principal financier, may appoint some members of the top management board, this board (as well as the management committees at different tiers of the system) will include elected representatives of users, the latter preferably constituting the majority of their membership.

(e) The bureaucracy, to implement the rules and regulations decided by the Board, should be employees of the autonomous authority responsible for each system.

(f) In order to reduce the burden on the bureaucracy administering the system, the responsibility of the system managers could be limited to providing water up to the outlet level. The system would enter into an explicit contract with each outlet users group specifying the volume, duration, and periodicity of supply as well as the amount each user group is expected to pay. The contract will specify penalties for non-compliance and, where appropriate, the terms on which such rights can be exchanged or traded with others.

(g) Each user group will be free to decide on how its supplies will be managed, the basis of allocation of water among members of the group, the crops to be grown, the basis for charging users, and method of collection.

Phased Strategy

The committee recognized that the proposed structure involves a drastic change in the existing regime and therefore would not be immediately feasible: Redefining allocation rules of existing systems, improving the quality of service, and laying down generally acceptable criteria for sharing shortages will take time. Also, effective control over water deliveries according to prescribed schedule is often not possible without substantial modifications in operational procedures, establishment of mechanisms for better monitoring of flows, and additional investments in communications and regulatory structures at appropriate points in the distribution network.

The changes envisaged involve a major reorientation in the roles and attitudes of the government (and its irrigation bureaucracy) and the cultivators. The former, being familiar with the existing arrangements, which have long been in vogue and which give officials considerable discretionary power (and the attendant opportunities for 'rent seeking'), are likely to resist any substantial reduction in their power and/or an increase in their accountability to users. The users, on the other hand, are unfamiliar with the intricacies of large system management and uncertain of the implications of a change in regime. They are also likely to be averse to shouldering the responsibility for mediating and settling conflicts among themselves. Both types of resistance cannot be worn down easily or soon.

It is therefore necessary to work out a phased strategy to change the structure and functioning of institutions concerned with canal management. A three phased transition has been proposed. The first phase, according to the Committee, should focus primarily on rationalizing the existing system of individual assessment based on irrigated area under different crops, to one of season-specific area rates reflecting differences in irrigation requirements between seasons and crops. This would make various crop rates correspond more closely to the volume of irrigation water they use. The level of cost recovery to be aimed at this phase is also quite modest.

A shift to a fully-fledged volumetric system could be made in the second phase after effecting the necessary physical changes in the distribution system for more effective regulation of water deliveries and reworking the operational rules and procedures for the main system. The aim should be to work out clearly defined criteria and procedures for water allocation based on a proper study of the actual conditions of current use, the relative merits (in terms of equity and productivity) of different patterns of use under different conditions of supply, and the pattern that is most acceptable to users. By bringing about a more assured and predictable supply of water between seasons (and within seasons), and leaving the farmers the flexibility to determine how best to use the water, the modifications would add substantially to productivity and may well lead to a saving of water which can be used to extend irrigation to a larger area.

Consultation with users is crucial to evolve an informed and acceptable set of operational rules for allocation. Ultimately however the state cannot escape the responsibility for striking a balance between competing claims of productivity and equity in distribution and of the conflicting claims of different segments of the system. This is likely to be more difficult in the case of existing systems (inasmuch as the claims that various users have established in practice will be affected) than in new systems where the planners have a cleaner state on which to work out allocation rules in consultation with potential uses.

Incentives and Pressure

In order to overcome the reluctance of farmers to take on the responsibility for group management, a system of incentives has been proposed, to strongly discourage individual service in favour of farmers' groups by making the revised rates, in the transitional phase, substantially less for those who opt for a group based volumetric levy, by giving preferential allocation of government funds to farmers' groups to enable better use of water and to award such groups contracts for maintenance of system facilities in their vicinity.

The Committee went on to argue the need to combine these 'incentives' with 'pressure' and 'positive measures to support and nurture groups'.

In order to exert pressure, the government must declare its intention to withdraw, after a designated period of 5–10 years, from the responsibilities for management below the outlet and confine itself to delivering water for a specified duration at the minors or the outlets. The message to the farming community should be clear that the government considers the water user groups as the main instrument for improving the management of the irrigation system. Government's commitment to the improvement of irrigation efficiency and farm productivity should also be visible, and the farmers should perceive the political will to improve cost recovery. This policy initiative should also be reflected in a time bound programme of introducing group delivery and volumetric pricing.

The positive measures include, besides educating farmers about the rationale of the new system and its advantages, the cultivation of a supportive attitude on the part of the departments concerned (including the Irrigation department) at all levels to the formation of groups, the provision of technical advice and assistance in working out rules and procedures for their operation, and the encouragement of voluntary organisations to play a larger role in the process. [GOI, PC, 1992: 131–2.]

The report was submitted 5 years ago and has been gathering dust along with many other reports in the archives of the Planning Commission. The Commission is yet to give a clear public indication of its position on the recommended reforms. The report is said to have been discussed with the State governments. The outcome of these discussion remains a well guarded secret. Beyond the rhetoric about encouraging user participation, there is little indication that the Commission is interested in pressing seriously for institutional reforms.

Given that irrigation makes a large difference to the productivity of land, and that supplies are short of demand, one would expect grass-root pressure for improving the management of irrigation systems. Complaints of inadequate and unreliable water supplies and their inequitable distribution are widespread. Substantial portions of the registered *ayacut* under major projects (usually at the tail end or the periphery of the command area) do not receive any water or at any rate adequate water, and at the same time other segments appropriate more than their entitlement. Illegal appropriation for irrigation of areas not falling within the command area by pumping water from rivers and canals, and other devices, is also not uncommon. Though those denied their share are numerous in terms of number and area cultivated, there are few instances of their exerting sustained and organized pressure on the government to press their claims and to check illegal appropriation of water.

One explanation is that the costs of this strategy are for too high in relation to the prospects of receiving redressal, and of the decisions being effectively implemented. The government has not been particularly consistent in observing the rules and regulations it has legislated. Its record in checking flagrant violations in spite of protests from the affected parties or even to enforce court decisions in cases of disputes is not impressive either. This dampens people's confidence in the prospects of success of, and hence incentives to, organize protests, political agitation, or litigation.

There are however signs of increasing farmer awareness of these issues and interest in organized effort at improving the quality of irrigation. The creation of water user associations, though confined to the outlet, has perhaps assisted this process. It however needs to be followed by a vigorous effort, which only NGOs can mount, at improving the level of water literacy, farmers understanding of the situation in specific systems, different ways of ensuring more predictable and fairer distribution, and their implications. Combined with credible initiatives on the part of the government to combine higher water rates along with measures to improve the management of the system, the prospects of significant institutional reforms can be substantially increased.

INTEGRATED WATERSHED DEVELOPMENT[11]

I have referred to the need to broaden the scope of the tank modernization programme to include treatment of the catchment areas, and of chains and networks of tanks. This is essential to check erosion, restore the vegetative cover in the upper reaches, and remove impediments to the free flow of water into the tanks. Together, in course of time, these measures are also expected to increase the proportion of local rainfall that is harnessed for local irrigation.

Watershed development has also a much wider role in improving the soil moisture regime on rain-fed land. A major reason for the low productivity of this land is soil erosion resulting in depletion of the topsoil cover, reduced capacity to trap and store *in situ* rainfall; and reduced soil fertility. Besides checking erosion, measures to increase the amount of rainfall that goes to build up soil moisture and various water harvesting devices, to harness (through percolation ponds and small tanks) rainfall for supplemental irrigation within micro catchments; and to increase the water supply to dryland crops and thereby create the conditions favourable to raising the yield.

In situ soil and moisture conservation works, provided they are properly conceived and implemented, would make a substantial difference to the level and pattern of water utilization. Illustratively, if the amount of water supply available to rain-fed arable land can be increased (through increased soil moisture storage and through water harvesting for local irrigation) by even as little as 5 cm per hectare, the effective utilization of water will increase by nearly 5 m. ha. m, which is about 10 per cent of the current effective use from all sources of irrigation.

The necessity for purposive measures to check soil erosion, improve the moisture retention capacity and the natural fertility of the soil, and reverse the progressive decline in the extent and quality of forest cover have of course been long recognized. Successive Five-year Plans have included a variety of programmes to this end, substantial resources have been allocated for them, and the scale of outlays has increased manifold. The approach has however been to tackle each of these problems separately and piecemeal rather than in an integrated manner. Thus the soil conservation programmes of the Agricultural departments and, subsequently, a variety of schemes aimed at raising the productivity of drylands have been confined to agricultural lands; construction and renovation of minor surface irrigation works is the responsibility of the PWD; while afforestation and social forestry are attended to by the Forest Department.

That contour bunding and gully plugging are not sufficient to check soil erosion and restore land to its full productive potential came to be explicitly recognized in the early sixties. The Fourth Plan in fact proposed 'an area saturation approach so as to treat all types of land on a complete watershed basis.' Master plans by basin were to be prepared to include afforestation, pasture development, terracing, and bunding of cultivated land, gully control and follow up measures to be coordinated among the concerned agencies. The concept of watershed development combined with improved techniques has been reiterated in subsequent plan documents, the emphasis being on small watersheds of 1000 to 2000 ha.

A series of integrated watershed development projects (other than those for upper catchments of major river valley projects and flood-prone rivers) were started during the seventies and early eighties. These include, (1) the ICAR operational research project for integrated development of 47 selected watersheds;[10] essentially meant as demonstration projects, all of them have been completed. (2) Projects for watershed development in rain-fed areas funded by the World Bank: four of these are currently under implementation. The average size of a watershed under this scheme is much larger (around 25,000 ha.) than that under the ICAR or the National Watershed Projects. (3) The National Watershed Management Project which, unlike the Bank project, focuses on micro watersheds (of average size less than 1000 ha.). This project, rechristened National Watershed Development Project for Rain-fed Areas (NWDPRA), is planned to cover all development blocks with less than 30 per cent of the area under irrigation.

These programmes provide for treatment to control soil erosion, optimum rainwater utilization, cropland improvement by using dry farming technology, farm forestry, horticulture, and support measures for training and research. While organizational arrangements vary, all are concerned to ensure coordinated action by the concerned departments in the project area. The more recent proposals also emphasize the necessity for consultation with, and participation of, the beneficiaries, as well as equitable distribution of benefits, but all of them see the government departments playing, at any rate in the immediate future, the key role.

Most programmes, including those such as the ICAR model watershed schemes and wasteland development under NWDB, which are supposed to be 'integrated', remain fragmented in terms of funding sources and responsibility for implementation. Thus the wastelands development programme draws funds from at least five different sources: The central government, the State plans, the Rural Employment Programmes, Drought Prone Area Programme (DPAP), and Desert Development Programme (DDP). Besides, different components of the model watershed plans are implemented by the respective line departments. Also conspicuously in evidence is a tendency for proliferation of various special area, rural development, and employment programmes, all of which have one component or another relevant to soil and moisture conservation, afforestation, and the like.

In Table 1 below one can see that there are nearly ten programmes dealing with one or more aspects relevant for watershed development. To the original list of departments involved in this activity, namely soil conservation, minor irrigation and forests, has been added another, the Department of Rural Development. The number of programmes under which each one of these activities are taken up has multiplied. The reporting mechanisms are so poor that it is difficult to say with confidence the degree to which a particular activity has been achieved under each programme. In every district the line departments pursue their own programmes largely unmindful of similar activity carried on by other departments. With rare exceptions (like W. Bengal) there is not even an attempt to avoid duplication of effort under different programmes.

A national workshop held in 1988 reviewed the experience of small-scale watershed schemes and came to the following conclusions. The watershed plans seem to be no more than a collection of departmental schemes. There is nothing like an integrated plan. Inter-department coordination has remained elusive: for example, the attempt of the Haryana

TABLE 1

Plan Programmes Under Which Different Elements of Watershed Development are Covered

Dept/Programme	Aspects Covered					
	Soil conservation	Land shaping/ development	Minor irrigation	Sylvipasture/ pasture	Social/ farm forestry	Afforestation
Agriculture						
Soil Conservation[1]	×	×	×	–	–	×
Integrated dryland agricultural development[2]	×	×	×	–	×	×
Special crop programmes[3]	?	?	×	–	–	–
Forests						
Afforestation[4]	–	–	–	×	×	×
Rural fuel-wood[5]	–	–	–	×	×	×
Soil watch[6]	×	×	–	–	×	–
Catchment area treated RVP, FPR[7]	×	–	–	×	×	×
Rural Development[8]						
DPAP	×	×	×	×	×	×
DDP	×	×	×	×	×	×
Rural employment programmes	–	×	–	×	–	–
National Wastelands Development Board[9]	×	–	–	×	×	×

1. Conventional soil conservation programmes cover only agricultural lands and concentrate on contour *bund*ing, gully plugging, *nallah* control, and other conservation works together with some afforestation of denuded areas.

2. Launched as pilot project in 1970–1 to test and demonstrate AICRPDA technology, modified in 1982 into integrated development of selected micro-watersheds and promoting available techniques of dryland farming. Covers soil conservation, land development, construction of water harvesting/storage structures, improving vegetation cover and improved seeds and fertilizers.

3. Special programmes for pulses, oilseeds, and cotton, which are largely dry crops, also provide for minor irrigation.

4. Includes seedlings for farm forestry.

(Contd.)

(Table 1 Contd.)

5. Rural fuel-wood plantation project introduced in Sixth Plan in 157 districts: fast growing trees, fodder and small timber in block plantations, and farm forestry.

6. Introduced in 1977–8 in selected micro-watersheds of the Himalayan States.

7. River valley project catchment projects covers 587 watersheds in 27 catchments. Flood-prone catchment projects cover 240 watesheds in 8 catchments of the Ganga basin.

8. DPAP covers practically all treatments figuring in watershed projects. Includes afforestation and pastures, explicitly on a watershed basis, with earmarked allocation.

DDP started in 1977–8, operates in 21 districts (17 hot, 4 cold centres) includes afforestation, pasture, shelter belt plantations, and sand-dune stabilization.

Rural employment programmes cover in principle all the categories of treatment figuring in watershed projects. However, social forestry and afforestation have earmarked allocations that are spent on projects of the NWDB, Forest Department, *zilla parishad*s, and cooperatives. The principal component of the wasteland development programme is to step up the rate of afforestation with people's participation. A shift of emphasis from social forestry and block plantations on government land to farm forestry and community plantation on private and community land.... Funds drawn from various other programme allocations.

9. NWDB set up in 1985 as a nodal, coordinating and monitoring body at the apex level.

and Punjab Governments to replicate the successful Sukhomajri/Nada models by creating coordination committees at various levels failed.

The State level committee was not able to meet frequently as the members had many other priorities. The district level committee met more frequently but was not very effective because the funds were not under its control. As the State level committee rarely met the coordination among the Government's Minor Irrigation, Forest, Soil Conservation and Revenue Departments was not very good. The Sukhomajri model was totally dependent on a sociological perspective which was not acceptable to the technical departments of the Government. [SPWD, 1989]

The workshop further noted that 'Lack of coordination and lack of integrated fund support go together... individual departments, each with its own limited schemes and funds, initiated unrelated projects in the same villages. There was no movement towards a common objective. Watershed development work taken up by the Government tended to suffer from bureaucratic ways like standardization and setting targets.' Technical expertise, so essential in demonstrating the value of the new approach to the people, was also lacking.[12]

The lessons of this experience are yet to make a significant impact on the way such projects are planned and implemented. The commitment to integrated watershed development has been reaffirmed and the scale of resources devoted to them have been expanded, but in reality 'integration' remains no more than a slogan. The newer generation of watershed projects (like CLUMP in Karnataka and the NWDPRA) continue to be cast in the mould of 'schematic' approach, based on norms for each of the major components, their scale, and unit cost. This may well be justified on grounds of convenience in presentation but one would like to see some recognition of the fact that the mix of treatments and unit costs are apt to vary from watershed to watershed; and that we do not quite know enough to be able to assert the magnitude and timing of the benefits with much confidence.

The organizational arrangements proposed are much the same as in the earlier programmes.[13] There is no indication of how the lacunae—in terms of the expertise available for the preparation of the plans, the quality of the plans, the efficacy of the State and district level councils and the watershed development teams as coordinating mechanisms or on the skill and motivation of the personnel—are to be remedied. Nor do they have much to say about how the community will be involved in the formulation of the plans, or how the 'optimum pattern of resource utilization' will be enforced on the beneficiaries individually and collectively.

There is much rhetoric about active involvement of beneficiaries in all phases of the project. 'Stimulating and promoting people's participation in project planning, project preparation, implementation, and post-project management of project assets will be an integral part of the approach and strategy' of the NWDPRA. This is to be achieved through the village *panchayat*s or through specific organizations to participate in all these plans, especially 'post-project maintenance and operation of community assets.' A significant role is visualized for NGOs in creating awareness, training and evaluation, and monitoring. Converting the watershed development project from a government scheme to a people's movement is the ultimate objective.

The guidelines speak of learning

a lot from the village community and unlearn some of their orthodox views and technical assumptions about people's capabilities. In the ultimate analysis science and technology from research institutes, technical and managerial know-how of the project staff and accumulated experience of the village community shall be systematically integrated.

Unexceptionable sentiments, but one is yet to see any concrete indication of the mechanisms of interaction between communities and the departments, between the departments concerned, and within communities on the basis of which we can look forward to improving the past record.

Programmes of NGOs[14]

Non-governmental and voluntary organizations have also shown a growing interest in integrated watershed development. Some, like Ralegon Siddhi and Jawaja, started out tackling a particular problem (like water management and wasteland development) and came to the realization that a broader, more inclusive, programme for ensuring optimum use of available land and water resources was necessary. Others (like Sukhomajri, Tejpura, Naigoan, and Daltonganj) were conceived as integrated watershed projects. There have been doubts whether the watershed or the village should be the unit of planning, but in either case the idea of a coordinated plan covering all land now commands widespread acceptance.

Non-governmental activity is quite limited in scale; and NGO projects are largely dependent on government funding. However, compared to the governmental programmes those involving non-governmental and voluntary organizations reveal an approach that is very different, more exploratory and open-minded. Marked by diverse social philosophies and approaches to local mobilization, they show a much greater awareness (a) of the need for flexibility in seeking technical and organizational solutions in the light of specific local conditions; (b) the tension between equitable distribution of the benefits from the development and the logic of local social and political configurations; (c) the problems in evoking community interest and participation and sustaining it; and (d) the need for innovative adaptations and the importance of learning from experience.

Many of these projects make a conscious effort to utilize local knowledge and experience, and to adapt programmes to circumstances that are highly variable. They have also shown a keen concern to evolve better, more cost-effective techniques (e.g. through proper choice of species, natural regeneration rather than of plantations, reducing costs of raising nurseries and improving the survival rates of planted saplings) for improving the effectiveness of investments. There are also instances of experimentation with new designs and materials to reduce costs of structures, evolving multi-tier crop-cum-tree systems to make more effective use of soil moisture and simultaneously augmenting the moisture storage capacity of the soil, and developing low cost water distribution systems using local materials and skills. (See, for instance, Datye, 1997.)

The NGOs are acutely conscious that such extensive local resource development programmes cannot succeed without the consent, involvement, and active participation of the beneficiary communities. Their effort to secure community involvement run into the problems created by the differing perceptions and conflicting interests of various groups. A great deal of time and effort goes into tackling these problems. One also notices a marked concern for equitable sharing of the benefits of the programme, particularly by the poor and disadvantaged groups. Some remarkable examples of success in tackling these problems have been reported, e.g. Sukhomajri, Ralegon Siddhi, Pani Panchayat of Salunke, to mention only a few. However, as the 1988 SPWD workshop showed, there are serious difficulties in securing people's involvement and ensuring equity; at any rate, there is no single general pattern. It is also apparent that the social orientation of the NGO groups is often much stronger than their technical skills/experience and that a great deal of their effort is spent in mobilizing funds and liaising with the concerned Government departments who are not always favourably disposed to such initiatives.

That integrated watershed development programme is an excellent and eminently rational concept is scarcely in doubt. Piecemeal treatments to address particular categories of land and particular aspects of the problem will not do. Soil conservation of farmlands or construction of irrigation ponds will be of limited value unless restoration of the vegetable cover on degraded upper slopes are also taken in hand and properly maintained. Appropriate treatments for all categories of land in the watershed is essential for lasting improvement in productivity and ecological stability. It is however obvious that the present approach to planning and implementation of this programme leaves a great deal to be desired. Besides improving the quality of information and technology inputs for the programme, drastic changes are needed in the way it is organized.

Database for Planning Watershed Development

Watershed planning demands a great deal of information on topography, vegetation, soils, present and potential use patterns and productivity of land resources in relation to each selected watershed. The inadequacy of the organization and skills necessary for collecting and interpreting the data has been widely noted. It would however take an inordinately long time and be far too expensive if we were to rely exclusively on organizations like ICAR research institutes, the All India Soil and Land

Use Survey, and the Survey of India. Speedier and less expensive techniques and organizations have to be evolved. A two-pronged approach is desirable.

First, much greater use should be made of remote sensing methods for mapping basic land forms, land use, vegetation and water resources on a micro-watershed basis. This is a powerful technique whose range and precision is constantly being improved. Interpretative capability exists in the NRSA, ISRO, space application centres in several states, as well as a number universities and research institutes. The NRSA has set up a special unit (NRDMS) to collect data needed for micro-level spatial planning.[15] The scale is however limited, the resources appear fragmented, and the problems of applying the technique on a large scale, including those of ground verification, have not been effectively resolved.

In order to harness the potential of remote sensing for watershed programmes it is essential to have a comprehensive review involving both the experts and the users of the data, to (a) assess the specific types of information relevant to watershed planning that this technique can generate; the kind of ground truth verification needed to firm up the information base; (b) the level of precision and costs; (c) the scale on which this activity can be undertaken with available personnel and equipment; and (d) the rate at which these capabilities can be augmented. The senseless restrictions on making the satellite imagery available to the public also need to be removed if the technique is to be of much use for watershed planning.

This still leaves a great deal of other data (for example, the status of land use contours; nature, depth, and fertility status of soils; degree of erosion in different parts of the watershed; extent, nature, and use of tree and grass species grown locally; distribution of land ownership; the extent of common land and its legal status) to be collected in each watershed. At present such data, to the extent that they are collected at all, are compiled wholly through the government bureaucracy. This is neither desirable (because of the time and cost involved), nor necessary.

Much of the information can be collected by reasonably educated members of each community after some training. Relatively simple and inexpensive techniques for contour mapping, permeability measurements, and soil analysis are available. A recent pilot experiment at the Centre for Earth Sciences in Kerala has shown that trained volunteers from the community can produce reliable and detailed maps of high quality with the help of cadastral maps and some supervision.[16] Such techniques, implicit in the idea of 'barefoot' ecology workers advocated by Anil Agarwal, should be seriously pursued. Apart from the saving in time and cost, it could be an extraordinarily effective way of educating people in each community about the basic concepts, techniques of measurement and their significance, and will prepare the ground for more informed local participation in formulating and implementing the actual programme.

It is, at the same time, essential to recognize that one cannot wait for all the relevant information before embarking on the projects. The information available, especially in the initial stages, is bound to be incomplete and even unreliable; nor is there sufficient experience in other similar contexts to fall back upon. Under these circumstances a certain amount of trial and error, learning from experience, and gradual step-by-step upgradation of the programmes is imperative and must be consciously built into the design of organizations for watershed planning.

Technology Inputs[17]

Technically, investment in watershed project involves the following categories of activity: (a) soil conservation, (b) restoration of tree/grass cover, (c) measures to improve soil moisture storage, and (d) percolation ponds and minor irrigation structures. Each of these involves a combination of several measures, and project planners must choose the right mix of measures and proper design (from the engineering viewpoint and in terms of cost effectiveness) of each.

Given the great diversity in terms of topography, terrain, agro-climatic conditions, extent of degradation and existing land-use patterns, the specific treatments and techniques cannot be reduced to a standard pattern to be applied uniformly. The kind of measures needed in the upper Himalayan catchments will obviously be very different from those in the Western or Eastern Ghats; the arid and semi-arid tracts of the Deccan and Rajasthan will call for different approaches from that say, in Chhotanagpur and much of central India; the problems of the Gangetic plain and the deltas are of an altogether different character. It would perhaps help in determining relative priorities and differentiating the appropriate treatments if watersheds were classified into distinct types in terms of the relevant characteristics. This again is recognized in principle, but the emphasis on schematic budgets and the poor mechanisms of monitoring make one doubt whether there is adequate sensitivity to these aspects. At any rate there is scarcely any systematic analysis of the experience of the government watershed programmes in this respect.

Considerable controversy exists on the appropriate type of structures, materials, and construction techniques to be used. There is a great deal to be said in favour of local materials; but not for relying wholly on traditional technology and skills. Tradition does not provide technologies to solve all problems; it certainly does not always provide the best solutions to problems involved in watershed development. It is unwise to deny the programme the benefits of different approaches and different technologies (including modern techniques) evolved elsewhere. A combination of local materials with modern engineering techniques may often be technically superior, cheaper, and also serve as a means of upgrading local skills with on the job training.

The limited experience so far has thrown up several interesting possibilities for innovative design using local timber and such other materials along with modern reinforcement techniques (in the form of geocrete, geomembranes, filter farbrics, ferrocement, etc.) which improves performance and/or cuts costs. Illustrative calculations of designs using alternative materials suggest potential cost savings ranging from 30 to 80 per cent, especially in the construction of small dams, embankments, ponds, and water distribution (Datye and Paranjape, 1988; and Datye, 1997).

Similar opportunities for innovation exist in the context of natural versus artificial regeneration of forests, choice of species for planting, techniques of raising nurseries, techniques for more economic and productive use of water. Several projects (especially those with which NGOs are involved) have consciously explored alternative cheaper solutions to these problems. The solutions, it may be noted, are not always technical. Often (as in the case of natural regeneration, nurseries, and tending saplings) they require institutional measures. Examples of technical innovation include the multi-tier cropping system and cheaper water conveyance systems. For instance, if crop and tree species are so chosen as to combine or alternate shallow and deep-rooted species, the volume available for storage of moisture in the soil and the effectiveness with which this storage is used for producing biomass is greatly enhanced even as the recycling of biomass could contribute significantly to improving the quality of the soil. The nature of the crop pattern and the water distribution system has a significant bearing on the quantum of additional output per unit of water and, therefore, on how widely the benefits can be shared.

We need not accept all these claims as valid to recognize that there are exciting possibilities of creative combination of traditional materials and skills with those of modern technology. Will this not make the watershed development dependent on outside expertise? Clearly engineering design, and developing/adapting newer materials or reinforcement techniques will require the support of trained technical personnel and institutions. But if they are brought into a continuing process of interaction with the beneficiaries, with good networking for dissemination of information regarding new technical developments and the experience of their actual application, there need be no danger of dependence. It should be possible to develop a mutually beneficial and creative relationship in which universities and research institutes can also be meaningfully involved.

Costs and Benefits

At present watershed development, like much else in rural development, is largely funded by the state. The quantum of resources available for such activity is very much larger than the amount properly spent on so called 'watershed projects'. A great deal of the expenditure is however, as already noted, spent on particular components of the activity that figures in watershed projects but in a totally piecemeal, un-coordinated way. By shifting to the integrated watershed approach, the available resources can be put to far more effective use, and that is by itself a very strong argument for such a shift.

Nevertheless, there is need for much closer scrutiny of costs in relation to expected benefits. At present, systematic analysis of these aspects is rare. There is scarcely any effort and, for that matter, even interest in expert scrutiny of alternative designs for different treatments, the content of treatments appropriate to different agro–ecological contexts, and their costs. *Ex post* studies comparing the original expectation of the performance costs and productivity impact of different interventions with what is actually realised are rare. Analysis of the reasons for divergence and their lessons for future projects is practically nonexistent.

Hardly any systematic baseline surveys are conducted of land use and production of crops, timber, fuel, fodder, and animal products before a project is initiated. Assessment reports of a few watershed projects give the estimated crop yields, usually for one year, compared to non-treated areas and occasionally pre-project yields. Estimates of fuel, fodder, and minor forest fodder collected by the beneficiary community before and after a watershed project are seldom available.[18] A proper assessment productivity impact must cover all these aspects by comparing the situation over a 3–4 year period before the project and after completion of the treatments. Single year data, being affected by the volatility of weather conditions, can be quite misleading. Ideally, these assessments have to be done for several years after the treatments are complete because the full benefits take time to manifest themselves and there is always the danger that the facilities created are not managed properly on a continuing basis.

Evidently the institutional arrangements for compiling such data on a scientific and objective basis are very weak, if they exist at all. It is obviously difficult, both in terms of organization and cost, to conduct such elaborate surveys in all watershed projects. Nor it is necessary to cover all watershed projects. The requirements of planners and the need to learn from experience would be amply met by independent monitoring both before the initiation of works, and periodically after the completion of treatments, of a select *sample* of projects typical of different situations. A continuing professional organization as an integral adjunct to the programme is essential. For the rest, each watershed community must be encouraged to make their own arrangements to collect data on the implementation aspects of the project and its impact in their respective localities. Educated people in each community with some training can undertake this work and help the community to review the results in an informed manner.

Programme Organization and Management

We have already seen that the watershed development programme is dominated by the bureaucracy with far too many departments being involved, and integration and planning being conspicuous by its absence. There is near unanimity about the desirability of bringing all the concerned activities under unified control at the watershed level. All the watershed development programmes have devised, on paper, arrangements to get the concerned departments to work together in achieving this goal. But clearly these arrangements have not succeeded in breaking down the parochialism of line departments. The proliferation of programmes under which watershed related activities are taken up has added to the confusion. Broadening the departmental outlook of the concerned department staff and getting them to view their particular segment in relation to other aspects of watershed development will need drastic action.

There is really no need for separate programmes for soil conservation, social forestry, and minor surface irrigation works, nor for separate area programmes like wastelands development, DDP, and DPAP. Much of the tribal area development and the various rural employment programmes can also be merged into local area development programmes centred around the watershed. Pooling all the resources under a single programme will help reduce the duplication, fragmentation, and the resulting waste that characterizes the existing dispensation. It is therefore rational to merge all local area development and employment programmes into single pool and transferred through the States on the basis of well-defined criteria, to the *zilla parishad*s and other elected local government institutions. A substantial part of this, the exact amounts to be determined in the light of local conditions, can then be used for watershed development as part of local area plans.

Second, and as a corollary, the field staff of concerned line departments at the district level and below should be brought under the unified control of a watershed project manager selected for his managerial rather than specialized technical skills. The higher echelons of the line departments will then be primarily responsible for promotion of research, collating the research findings as well as experiences gained from projects all over; organizing independent technical assessments of the design and performance of works done in various projects and disseminating them to the watershed project managers as well as people at large; and maintaining high standards of the technical personnel through training. The senior personnel of line departments could also play a very useful role in periodic supervision of the implementation of the area specific watershed projects (Bali, 1988).[19]

Such an arrangement would obviously be quite a radical departure from existing patterns. Administrators do not react favourably to the idea. However, the experience of the committee system of coordination in rural watershed programmes as well as the Command Area Programme, which was another failed attempt to achieve integration, make such a departure inescapable if we are really serious about integrated watershed development.

Ideally both the state agencies and the NGOs should play the role of catalysts in stimulating village communities' interest in watershed development and in providing them the necessary technical and financial support. Such an approach cannot however be effective unless its benefits are convincingly demonstrated on a sufficiently large scale and in diverse situations. Also, given the weakness of democratic institutions of local government and the fact that village communities are often segmented by caste, landholding, and other factors, no single pattern may be feasible. It is important to recognize that we are still in an experimental stage, both technically and institutionally and, therefore, ample scope should be provided for experimentation with different approaches.

It is useful, in this context, to distinguish between three types of situations: (1) Those watersheds that consist predominantly of forests or which are situated in the upper reaches and higher elevations of river catchments and where soil conservation and afforestation will be the principal component of watershed treatment. (2) Those where a substantial part of the area is, or should be, under forests but a sizeable human population depends in various ways on the forests for their living (most tribal areas fall under this category). (3) All the rest, where a large part of the land is under settled cultivation. The range of treatments required, as well as the nature of the technical and institutional problems to be resolved, in these three situations are different.

The first category can be largely left to the forest departments. The necessary complement of expertise from the other relevant disciplines (soil conservation, civil engineering) can be inducted into the department. In the second category, which is also likely to be in predominantly tribal areas, thé Forest Department will clearly have a major role. Planning is likely to be more complex inasmuch as issues regarding the rights of the tribal population in relation to forests, the optimum development of land use to meet their needs consistent with ecological balance, and ensuring that desirable land use is actually achieved and sustained have to be addressed. In the third category the changes in the land use pattern and soil conservation measures will have to be primarily concerned with achieving the maximum sustainable level of productivity (in terms of food, fibre, and timber) to meet the needs of the population dependent on the watershed.

Community Participation[20]

Consolidating all local land, water, and forest programme area development projects with the watershed as the basic unit, and bringing the government personnel involved in this programme under a unified command at the field level are necessary but not sufficient. The watershed teams should be prepared not only to carry out projects on their own, but also to help NGOs and village communities who may be interested in doing so. In all cases the state agencies should actively explore, (a) ways in which local people's knowledge, experience, and sense of priorities can be combined with the available scientific and technological knowledge to mount a more meaningful and therefore more effective programme; and (b) ways in which a credible (i.e. enforceable) collective consensus on the content of the programme, its management, and sharing of benefits can be forged.

There exists a broad consensus on the desirability of people's involvement in watershed programmes. In reality, however, even consultation with the beneficiary communities is minimal, let alone any meaningful participation in any phase of the programme. The government bureaucracy is simply not used to the idea and is frankly quite sceptical whether it will work at all. (They are not, it should be said in fairness, the only ones to be sceptical) Many reasons are cited: illiterate villagers cannot grasp the technical problems of engineering and managing such a complex activity; the communities are so divided and full of conflict that they can seldom agree on a common programme; and handing over large sums to such communities will either lead to waste or be used to the advantage of the already better off and powerful segments of the rural communities. Similar arguments have been made against democratic decentralization.

Some of this, especially questioning the relevance of villagers' knowledge concerning their immediate environment for planning local development works and the notion that villagers will use resource more wastefully and less equitably than the state agencies, reflects prejudices rather than established facts. In any event, the projects prepared by the state agencies are themselves less than satisfactory partly because of insufficiency of expertise and motivation among the officials and, to an important degree, due to officials' unfamiliarity with the specific local conditions and problems. Having said this, one should not also make the mistake of romanticizing the beneficiary communities' capacity to conceptualize what it takes to develop the local resources; work out the appropriate technical solutions and their implications, and understand how different facets (technical and social) of this process relate to one another. Nor can anyone wish away the existence of conflicts in village communities.

Involvement of external agencies (both governmental and non-governmental) is essential to make the community appreciate the necessity for watershed development and its implications; to bring in the relevant knowledge to tackle the problem; and, where necessary, to safeguard the interests of the weaker and poorer segments of society. This last mentioned intervention is needed especially when democratically elected representative institutions of local government are yet in a fragile, fledgling, state and the poor are not in a position to press their interests effectively. Since the bulk of the funding of the watershed programme is provided by the state, the government agencies can legitimately intervene to protect the interests of the poor. The question is what kind of a role these agencies should play.

That this can be done is amply borne out by the experience of several NGOs which have persisted in getting the community involved without being daunted by failures. It may be unrealistic to expect such dedication in a large bureaucracy, but the state bureaucracy working in collaboration with NGOs could make a significant difference, provided there is an acceptance of their complementarity and a mutually supportive relationship is established between them.

While this takes effort on the part of both the state and the NGOs, the state agencies have perhaps to exert much more to change their long entrenched attitudes towards non-governmental initiatives. One implication of this is to permit a great deal of latitude in deciding the scope and content of particular area programmes, for experimentation, and gradual learning from experience. Schematic budgets, rigid specification of scope and content, insistence on targets and 'norms' for unit costs are all inimical to such an approach.

The problem of securing an enforceable collective consensus is considerably more difficult. What is good for the community and what equitable distribution of benefit should be are matters that cannot be decided by an external agency. At any rate, such a fiat will not be enforceable; it will be frustrated in numerous ways if the centres of local power do not find it acceptable. A group of people will agree to cooperate for the collective good provided, (a) such action is expected to bring substantial benefits to the group as a whole both in relation with the size of group and to the costs involved; (b) the sharing of costs and extra benefits resulting from such collective action is seen by the group to be 'fair'; and (c) there are credible arrangements for ensuring that the agreed sharing of costs and benefits will be effectively enforced.

In the case of watershed development, the magnitude of likely benefits and their time-phasing cannot be predicted, in the current state of knowledge, with a reasonable degree of precision, even by experts. This being the case, it would not be easy to convince potential beneficiaries that their acceptance of the obligations in collective projects is worth their while. In this context, the state and NGOs have an important role in demonstrating that well-planned, integrated watershed projects can indeed make a substantial increase in the overall productivity of land.

Even if the state were to meet the bulk of the initial investment cost, the members of the beneficiary community have to accept obligations for proper maintenance of *bunds* and other physical structures, caring for the trees, and ensuring that the rate of cutting of timber and grass is regulated within well-defined limits. This entails restraints on the existing pattern of access and use of common resources, and also contributions for maintaining the assets and the organization to manage them. The willingness to accept these depends critically on how the participants view the 'fairness' of the burden sharing in relation to the benefit sharing (including dealing effectively with the so-called free riders).

Where, as is typically the case in Indian villages, there is no prior experience with institutional arrangements for development and regulation of common production resources, peoples' assessment is likely to be strongly coloured by their experience of the way its community institutions have functioned in the past. That these institutions have not been 'fair' to the lower castes and to those without any asset base will surely affect attitudes to fresh collective efforts, especially in relatively uncharted areas. In view of all this it would seem worthwhile for the government and the NGOs to make a special effort to select areas for demonstration projects where the potential for intra-community conflicts is low. This is apt to be the case when the community is marked by relatively low class–caste differentiation and/or a strong, generally respected, local leadership.

The uncertainties about benefits, and conflicts over their sharing, are compounded by the present legal position regarding the rights of government, the village community, and individuals over land which is *not* under private ownership. One problem relates to forests: ownership of all forest land is by law vested with the state. Prior to the coming of the British, the forests were an important source of livelihood for those living in and around them. This 'constructive dependence' of people on forests began to weaken when the British declared forests, which were traditionally controlled and regulated by the local communities, as state property. Indiscriminate exploitation for timber, and the spread of plantation forestry have reduced forests area and the availability of minor forest produce to forest dwellers. The traditional rights of tribals to collect a variety of produce in protected forests have been restricted by reclassifying large areas as reserved forests (Gadgil, 1989; Fernandez, 1989). Under these conditions the beneficiary population is unlikely to have much interest or stake in participating in watershed development.

A similar problem also exists in non-tribal areas. Data on the distribution of land by legal title is hard to come by. A recent survey in selected districts from different parts of India suggests that the ownership of 'idle' land is distributed between the cultivators, village *panchayat,* and the Government. The proportion owned by cultivators (i.e. private property) varies from as little as 17 per cent to 100 per cent; that owned by *panchayat* from nothing to 58 per cent, and Government-owned land between nil and 48 per cent (Raheja and Banerjee, 1988). The ownership of uncultivated land is thus fragmented: part of it may be already encroached upon and effectively under private control. The development and access to forest land is controlled by the Forest Department which is quite reluctant to give up its control to the village community. The status of village commons, to the extent that they have not been appropriated by private individuals, are ill-defined and the rights of exploitation are not clearly codified.

These anomalies and confusion can be substantially mitigated by the following measures: The Forest Department should be made solely responsible for protecting 'areas which are genetically rich and ecologically fragile and which must... have an undisturbed natural forest cover' (Agarwal and Narain, 1989), and also areas, such as in upper catchments of rivers and in hilly tracts, which are (or should be) mostly under forest cover. In the former, 'mining and all the other such human activity banned or extremely restricted.' However, regulated usufructary rights to fuel, fodder, medicinal herbs, and other produce essential to people living in and around the forests should be permitted/restored. This is especially important in tribal areas.

All other degraded and uncultivated land, which is now 'owned' by the government or the village community, should be brought under the control of village communities taking care to clearly demarcate the boundaries of common land vesting with each village, making it legally obligatory for them to use such land for growing trees for the common benefit of its members. It has been suggested that the *panchayat* should be legally accountable for negligence in discharging this responsibility and that, to this end, the maintenance and regeneration of forests on common lands should be made the first charge on revenues earned from such land. All this requires thorough review and recasting of laws relating to common land, forests, forest conservation, and the rights of individuals to the produce of such land (Agarwal and Narain, 1989; Chattrapathi Singh, 1988).

There are also legal impediments to local initiatives in constructing small irrigation and water harvesting works (Agarwal and Narain, 1989; Datye, personal discussion). Under the existing law all rivers, including streams and rivulets, are government property. No individual or village community can use them to store water without the permission of the Government. There are several documented instances of objections from the PWD or its refusal to grant permission, thereby preventing the construction of such small storage or diversion structures by community effort. While the Government does have a responsibility to ensure that the available water in a river basin is optimally utilized and in such a way as to distribute the benefits in a reasonable way to the various parts of the basin, the present restrictions on small, purely local, works should be substantially reduced in the interest of promoting watershed development.

Distributional Aspects

A clearer, simpler enunciation of legal rights in relation to common land and local water conservation works will clearly help. There however remains the problem of arriving at an agreement within the community relating to the sharing of costs and benefits, and of devising mechanisms by which this sharing can be enforced. Some have taken the view, on the basis of experience of projects like Pani *panchayat* and Sukhomajri, that a strictly egalitarian sharing of the benefits arising from the development of common land is not only desirable but essential to generate peoples' participation and contribution. There is reason to doubt the wisdom of elevating this to a universal principle.

In Sukhomajri, for instance, everyone had an equal share in water. Those who did not have land sometimes leased in land in exchange for their water rights; some sold their share of water to others with land. In the nearby village Dhamala there was, however, not even an attempt at egalitarian distribution: only the landed got water in proportion to their landholding, but this did not lead to the society falling apart. Complete equality is thus not necessary for the success of cooperative effort. Viable collective institutions are feasible provided all parties are persuaded that the new arrangement will make them better off in some way and that rules regarding the maintenance and use of common resources are properly defined and enforced. The quality of local leadership is, therefore, a critical factor (Chopra et al., 1990).

Given that village society is divided by caste and by command over productive assets, the sharing of which a particular community considers reasonably fair for all concerned may fall short of the ideal. The acceptable and enforceable compromise will necessarily reflect the extent of social and economic inequality in each community, the level of general poverty/deprivation, and the character of the leadership. If a compromise needs to be made between the workable and the ideal, the choice should perhaps tilt towards the workable, for without that the prospects of raising the overall productivity of the watershed's land and water resources may be jeopardized.

At the same time, we need to incorporate in the project ways in which a fairer distribution of benefits and costs in favour of the poor/vulnerable segments can be promoted. First and most important, is an insistance that all aspects of the issue be openly discussed in the community before the watershed plan is drawn up. The necessity of giving the poorer segments a sizeable share of extra fuel, fodder, and water generated by the project could be explicitly made as one of the conditions of state assistance to the project. The Pani *Panchayat* technique of giving an equal share in available water (irrespective of whether or not a person owns land) is one way. Reserving the benefits of wage employment during the developmental stage exclusively for the disadvantaged groups, and leasing of forest and grassland to be maintained by the poorer groups in exchange for rights to exploit and market the produce are some others. In some cases the poor have been given access to certain designated quantities of fuel and fodder. As far as possible, sharing rules should be such as to create incentives for their observance by individuals and groups so that reliance on enforcement by fiat is minimized. Relying purely on fiat is in any case unlikely to be effective.

Requiring individual beneficiaries to meet all or a substantial part of the costs of improvements effected on their private land, with the state or the community bearing the cost of improvements to common land, clearly helps equity. Yet another technique, which is reported to have been fairly successful in Sukhomajri and Daltonganj is to require a substantial part of the incremental produce for further investment in community assets. A variety of such devices have evidently been tried out, especially in projects in which NGOs have been active. Government projects have been far less sensitive to this aspect of practice, if not in concept. A closer study of the experience of successful NGO-managed projects will give valuable clues in improving the organizational design of watershed projects.

REGULATING GROUNDWATER USE[21]

One of the most striking features of irrigation development in India during the past four decades is the rapid growth in the use of groundwater.

The expansion in groundwater irrigation is the result of several factors: improvements in the technology of drilling and lifting water have made it possible to tap deeper aquifers, pump larger volumes of water, and lowered the cost per unit of water pumped. The rural electrification programme, which has been largely funded by the state, together with liberal loan assistance on concessional terms for setting up, deepening and energizing wells and subsidized supply of energy (especially electricity) have contained the increase in private costs. The advent of new varieties and fertilizers increased the productivity impact of irrigation and the prices of agricultural products have been rising. The productivity per unit of water tapped is much higher in the case of groundwater in comparison to surface irrigation because, (a) groundwater irrigation involves much less waste by way of conveyance and application losses, and (b) farmers have much greater flexibility to adjust timing and the quantum of water application to crop needs.

On all these grounds the government's policy of supporting and encouraging private groundwater development was, till recently, widely acclaimed. There is now, however, a growing apprehension that these policies may lead, and in some areas may have already led, to over-exploitation of groundwater. Evidence of a progressive decline in groundwater tables in several parts of the country is accumulating (Parihar et al., 1990; Bhatia, 1992; Sivanappan et al., 1987; Shah, 1993; Dhawan, 1995a; Janakarajan, 1996; Janakarajan and Vaidyanathan, 1998).[22] According to the Central and the state groundwater organizations, the intensity of groundwater exploitation has reached, or exceeded, sustainable levels in an increasing proportion of the development blocks throughout the country.

Admittedly, declining water table per se and the increase in the number of 'grey' and 'dark' blocks are a rather crude and not always reliable basis on which to judge whether a particular region has overshot the limits of sustainable use. Ideally we need to know the location and boundaries of aquifers, their respective average annual recharge and extraction rates. None of this information is readily available or has even been compiled. Estimates of annual recharge for different blocks based on empirical formulae are scarcely reliable, if only because the boundaries of administrative units (such as blocks or district) seldom coincide with configurations of sub-surface aquifers.

A progressive lowering of groundwater table by itself may not always indicate over-exploitation, in so far as farmers change their crop patterns and irrigation practices to make better use of available water. In order to make more confident assessment of local situations, we need not only data on the behaviour of the groundwater table, but also information on the volume of water extracted, the proportion of wells that are going dry and out of use, and the total area irrigated. Nevertheless, a large, progressive decline in the groundwater level is a cautionary signal which should be taken seriously and investigated in detail.

Typically, the decline in the water table is associated with an increase in the well density and a progressive increase in the depth of wells. This has quite important economic and social consequences. As the wells tapping an aquifer increase in number and become deeper, the volume of water extracted is unlikely to increase in the same proportion as the number of wells. After a point, the yield per well declines and then the investments incurred per unit of water lifted increases. The operating costs per unit of water (mainly energy) also increases because it has to be lifted over a greater height. Altogether, the costs per unit of water pumped increases both for individual owners and for groundwater users taken as a whole. The capacity to mobilize the necessary resources for deepening and pumping is far from equal between farmers. Those (typically small farmers) who cannot afford the extra investments may be forced to give up well irrigation and thereby lose out to the better endowed farms.[23]

One would have thought that, under these conditions, there would be a strong economic incentive, both for individuals and for society, to explore arrangements to contain the tendency for competitive deepening. Fewer wells would mean less expensive water for everyone provided arrangements (involving joint ownership and operation, community management, or a system of trading in water) for distributing the water among interested users are set in place. While evidence on these aspects is scanty, the general impression of persons knowledgeable in ground-water management is that such cost-saving institutional arrangements have not developed on a widespread or significant scale. Community management of groundwater is rare.[24] Marketing groundwater has emerged and spread with the growth of well irrigation, but very unevenly, and evidence suggests that the proportion of water sales to water extracted tend to be generally quite small.

The strong preference for individual ownership despite the higher costs involved, on the one hand suggests that individual exploitation even at higher costs is sufficiently productive to be economical. Individuals may also be prepared to bear higher costs because of difficulties in ensuring effective joint ownership and management of wells, and risks of depending on purchases from other well owners. Where the available supplies are inadequate to meet the needs of the area served by an aquifer, these constraints become more severe.

The social consequences of these developments are a source of much concern. Not everyone can afford his or her own well or to deepen the well to maintain access to well water in the context of a falling water table. Besides, in the absence of a credible collective institution or a widespread water market, such people (typically small cultivators) cannot hope to access well water at all, or risk losing such access as they may have had earlier. Competitive deepening therefore makes the distribution of access to groundwater increasingly skewed in favour of large, resource rich farmers leaving the small farmers at an increasing disadvantage in sharing the benefits from well irrigation. The concern about these adverse distribution consequences of the observed trends in groundwater irrigation and the necessity to mitigate, if not reverse, them finds strong expression in current debates on the subject. (See Dhawan, 1995; Saleth, 1996; Shah, 1993; Monech et al., 1995.)

There is one school that believes that efficient utilization of groundwater is best achieved through a system of competitive water markets, without any subsidization of capital or current inputs. The development of such markets is supposed to be facilitated by according legally sanctioned private property rights over, and the freedom to trade in, groundwater. There are however serious practical problems in implementing this approach; and its social consequences are very much a matter of contention.

In the first place, we have to contend with the inherent difficulties of legislating clearly defined, and enforceable, property rights in groundwater. At present the owner of a piece of land also has the right to exploit water (or any other resource) beneath it. Rarely however do the boundaries of a block of land owned by an individual coincide with the boundaries of the aquifer from which it draws groundwater. The latter typically extends over a much wider area than the former. Consequently, several (and often a large number of) people already have the legal right to tap a common aquifer. But under these conditions it is difficult, indeed impossible, to unambiguously define the property right in groundwater on an *individual* basis.

When many landowners have easement rights over an aquifer, the extent of water which any one of them can appropriate, and the costs involved, are, after a point, very dependent on how much and through what means others extract water. Thus, if A deepens his well and draws more water, his neighbour B may get less water and perhaps none at all! Legally the interference caused by A's action on B's water supply is justiciable. The aggrieved party (B) can go to court and get redress. Apart from the costs (time and resources) involved in getting redress, it is difficult to see the basis on which the court will decide the relative claims; and how its decisions will get enforced on the ground. The task *may* be tractable in a situation (like California) where relatively few farms tap from a given aquifer. When, as is the case in India, the number of farmers involved is large, the dispute potential becomes very high and resolution of disputes an unmanageably costly and complex process.

In principle, the rights of ownership and/or use of groundwater can be meaningfully defined over an aquifer as a whole and vested with a well defined entity that could be a privately owned enterprise (which wants to exploit the resource for its profit) or with individual communities of people who happen to own land on the surface above the aquifer, or even with all communities who happen to be located above the aquifer. The problem here is who will decide the basis on which the quantum and the terms of access are decided and how they will be enforced.

The proponents of the markets argue that by allowing whoever has been given the right over groundwater to extract it and use it to maximize profits, there will be strong incentives for keeping down the costs of extraction and to allocate the water among competing uses efficiently. Something like this seems to have happened in Khera district of Gujarat (for details see Shah 1993, 1997). But where, and this is the more typical situation in the country, the groundwater supply is inadequate to meet the demands and its economic value is high, purely private exploitation would lead either to monopolistic exploitation or to highly skewed distribution of the benefits in favour of the better-off farmers/users and at relatively high social cost. It implies accepting the fact that those who have resources will corner the benefits of groundwater and those who do not will get no groundwater or get more volatile and costly supplies. In the absence of an effective regulatory system, we will necessarily end up in this inequitable situation. There is reason to believe that this is in fact happening in several parts of the country.

Even in a situation like Khera, the market cannot control the tendency to over-exploit the resource in the interests of short-term private profit without any concern for long-term sustainability. There is also the problem that in areas like the Gangetic plain, where the aquifer spreads over an extensive area, the territorial unit over which groundwater rights can be meaningfully defined becomes unmanageably large.

If the market solution is unfeasible and undesirable, regulation by government bureaucracy alone is not a credible alternative either. In point of fact, over the past forty years, the government has sought, through enactments and administrative orders, to ensure 'equitable and sustainable' exploitation of groundwater. These are not conspicious for their success. A couple of examples show the problems involved.

In most surface irrigation projects in Tamil Nadu there used to be restrictions on digging wells in the canal command area. Fairly severe penalties were prescribed for violating this. In the Lower Bhavani project, for example, wells were prohibited within a certain distance of main/ branch canals. Soon after the project was commissioned, however, the *ayacutdar*s began digging wells on an extensive scale in contravention of the regulations. Attempts on the part of the project administrators to enforce penalties (some did in fact try to enforce these) were largely unsuccessful. The number of violations was so large, and political pressure on the government became so strong, that rule enforcement was soon abandoned (GOI, PC, 1964).

The government has also evolved a system of classifying development blocks into three categories, (1) the 'dark' blocks where the level of exploitation has reached or exceeded sustainable levels and where further development should be actively discouraged; (2) the 'grey blocks' with modest potential for further expansion; and (3) the 'white blocks' with considerable unexploited potential which should therefore receive support and encouragement from the government. There are of course questions about the validity of this classification.

The boundaries of the basic unit of this classification, namely the community development block, rarely coincide with those of hydrological units. The hydrological unit appropriate for such measurement is not amenable to standardized definitions. Thus, in no part of the contiguous and highly interconnected aquifers in the Indo-Gangetic plain is it really possible to talk meaningfully about sustainable rates of extraction. On the other hand, in the hard rock areas where aquifers are confined or isolated, river basins or sub-basins may not be meaningful units for assessing sustainable levels of use.

A second problem is of course that we do not have adequate and reliable measurements of either recharge or of extraction. Estimates of extraction are made from extremely patchy data. Measurement of actual extraction are rare. The Central Ground Water Board did some work in the seventies, relating extraction rates to energy consumption on the basis of some sample observations. We do not however know whether they were based on a properly designed survey in all parts of the country. Even if they were, they cannot be used to estimate extraction rates indefinitely because yield per well, depth, and therefore energy used per unit water are changing. In many parts of the country electricity use for irrigation is poorly metered or not measured at all. Nevertheless they are the best available basis for regulation at present. It is always possible to improve the information base.

The government has had several levers to regulate development of well irrigation: prospective well diggers have in several areas to obtain government permission; more importantly, government has control over financial institutions and over sanctioning power connections. Since the vast majority of wells and pumps are financed by loans from cooperatives and nationalized banks, and since the installation of electric pumps would not be possible without a power connection, the state has powerful means to check over-exploitation.

The state has however failed to use these instruments effectively everywhere, partly because of the lack of reliable information from the ground level and the difficulty of centralized monitoring of the accuracy of reporting and enforcement of rules by the lower level of the bureaucracy. On the other hand, by following a policy of subsidized credit and electricity, it increased the private profitability of well irrigation which in turn further increased the demand for more wells, deeper wells, and energization. Unable to resist these pressures, State governments all over India have been lax in regulating groundwater development in the interests of equity and sustainability.

Neither market mediated private exploitation nor wholly centralized regulation by government agencies appear to be workable or desirable alternatives. In this context the third alternative is to vest the right to exploit and allocate groundwater with local communities with the state confining itself to defining the framework of principles and procedures for community management and facilitating their functioning. Leaving the user communities to decide the basis for and mechanisms of exploitation and distribution of benefits may not however ensure equity or sustainability.

The working of community organizations in south Indian tanks (Vaidyanathan and Janakarajan, 1989; Rajagopal, 1991; Sivasubramaniam, 1997; Vaidyanathan (ed.), 1998) is conditioned by the configuration of power and distribution of resources within the communities. Since these configurations are unequal, one cannot expect that the distribution of such a scarce and valuable resource as groundwater will be equitable even in terms of ensuring access proportional to area held, leave alone tilting the distribution in favour of relatively worse off users.

Given the difficulty of correlating user communities with those of the aquifers, the problem of regulating the sharing of supplies from contiguous or interconnected aquifers covering extensive areas among the numerous communities served by them, will remain. The principal merit of involving user communities is to decentralize the task of monitoring the levels and patterns of use of available supplies as well as changes in water yields of wells and to evolve mechanisms for 'fair' distribution in the light of local circumstances. This reduces but does not eliminate the need for government intervention.

The state can and should do several things in order to facilitate sustainable exploitation. In the first place it needs to create strong incentives to check over-exploitation. The present policies of providing cheap credit, subsidized diesel oil, and near free electricity for agricultural use together encourage indiscriminate exploitation of groundwater. Withdrawal of these subsidies and charging farmers the real cost of providing these inputs for groundwater pumping is of crucial importance. Besides being desirable for restoration of the general fiscal health of the government, it would raise the cost of groundwater and induce a reduction in groundwater extraction as well as more careful use of this resource.[25]

This is the most important instrument in the hands of the government and its effective use is a necessary condition to check over-exploitation. But this is not a sufficient condition. The state also needs to regulate construction of new wells and deepening of existing wells in areas where groundwater is scanty, encourage community-based institutions to regulate the use of groundwater, improve the basis of assessing the current rates of exploitation, monitor more closely the signs of over-exploitation, and educate public opinion about the need for regulations.

It is imperative that the government play an active role in collecting information on the status of groundwater use, identify areas threatened by overuse, and especially highlight the experience of areas that have not heeded the warning and therefore suffered serious loss of output and income. The groundwater organizations should also review the present arrangements for monitoring groundwater to ensure that the choice of observation wells, both in terms of the number and their representative character, is adequate to provide a reliable picture.

In doing so, we need to draw a distinction between wells located in the command of surface irrigated areas and those that serve as the sole source of irrigation. Conjoint use increases the efficiency of use of water tapped by the surface system in the command. The spread of conjoint use in a particular surface system may, however, have consequences for the distribution of benefits both between different parts of the system and between that system and others downstream. The more we recycle the seepage in the upstream commands, the less may be available by way of regeneration flows of surface water downstream. Also, within a given command access may be unequal not only because all farmers may not be able to afford the resources needed to exploit groundwater but because of variation in sub-surface geology and the possibility that heavy pumping at the head of the command could affect both surface and groundwater supplies in the tail reach (see Vaidyanathan and Janakarajan, 1989; Janakarajan and Vaidyanathan, 1998). Evidently, assessing the impact of conjoint use on equity in distribution is far from simple.

In the case of pure groundwater areas, falling water levels is a warning signal but we need to know more about the consequences. Take the example of Coimbatore district in Tamil Nadu. The government and the farmers of the areas have witnessed a progressive increase in the number of wells, a lowering of the groundwater level, and increasing depth of wells for nearly four decades. However, official statistics do not show any reduction in the total area served by wells. This would suggest that access to well water is more widely dispersed among farmers, and though water available per well has declined they have adapted by changing their crop patterns and irrigation practices to conserve water.

In both categories of well irrigation, arrangements have to be made to measure extraction rates and alongside this to monitor groundwater table. Since these measurements are difficult and expensive, it is important to explore, (a) the relative merits of sample surveys of farmers as against measurements of actual pumping by independent investigation; (b) ways of estimating reliable 'norms' for different types of wells/regions or per unit of energy consumption and updating them; (c) the use of relatively inexpensive measuring devices that can be fitted to selected pumps and readings recorded at periodic intervals.

Our ability to sense the dangers and anticipate the pace of depletion of groundwater resources would be improved if there were systematic and continuous monitoring of, (i) the average yield per well, (ii) the proportion of wells going dry, (iii) the area irrigated per well, and (iv) cropping intensity and crop pattern in well irrigated areas. If any one or more of these dimensions show a deterioration, it would be possible to at least inform the population affected, alert them to the implications and the need for collective action to regulate extraction and ensure equitable distribution.

While active support and involvement of official agencies is essential, it is neither necessary not desirable that only these agencies should be charged with making the observations. The potential exists for using farmers and local institutions to monitor groundwater on a much larger, more systematic, scientifically satisfactory, sampling basis. The involvement of independent professionals in design, survey, and analysis of these data is also desirable. In any event, all the information should be placed in the public domain both to improve design, analysis, and interpretation at the expert level, and to make the findings and their implications widely known to the people in the concerned areas.

This is of the utmost importance in any effort to counter the myopia characteristic of both individuals and communities in matters relating to exploitation of natural resources. Like other processes having to do with ecology and the environment, the consequences of over exploitation are in the future; often the distant future. Even when people see the signals, and recognize the prospect of irreversible damage, reactions tend to be myopic. In north Gujarat, Saurashtra, for instance, water levels in tubewells are falling, rapidly. Even some of the more knowledgeable and thoughtful people in that area argue along the following lines:

Of course something should be done to arrest the depletion and to augment water availability through such means as artificial recharge. But since this cannot be done by individuals, people will keep drilling deeper and get as much out of this resource as they can immediately. May be 20 years from now this place will be a desert. But 20 years is a long time. Here and now is more important.

If knowledgeable individuals think like this, the prospect of mobilizing collective action to correct the situation is very problematic. There is no incentive for people to focus on the fact that may be twenty years from now things will become irreversible, that the present pattern is unsustainable, and that the area may become desert! Under these circumstances, even if the state sees the danger, and legislates regulations, the users' myopia prevents regulations from being implemented. Since it seems impossible to devise any incentive schemes by which the current generation can be made to care for the next generation, the only way is to very forcefully, and continuously, din into the public ears the emerging trends and their consequences.

There are two ways of doing this. One is to keep monitoring the water table, the proportion of dry wells, the yield per well, and the total area benefiting from well irrigation in all groundwater areas, particularly those that are over-exploited and those that show signs of progressive depletion. A complementary approach would be to dramatize the potential dangers, by highlighting the experience of areas that have neglected the signals and have come to grief. General talk about the process of degradation or depletion of groundwater will not do. One has to cite concrete examples of the experience of specific areas like Saurashtra and parts of Tamil Nadu where this process has gone unchecked and has done irredeemable damage. Social myopia can be countered only by publicizing and dinning into the public's ears the experience of such areas, their consequences and the extreme importance of taking collective action to regulate use in the long-term interests of the community and its succeeding generations.

NOTES

1. This section drew heavily on Vaidyanathan (1991, 1994a, b).

2. A systems model for determining optimal water allocations in the Bhakra Nangal system was built and estimated by researchers in the Indian Statistical Institute during the late sixties (Minhas et al., 1972). A joint Indo-US team also explored the possibilities, but in more general terms, during the early 1970s (GOI, 1972). See also Chaturvedi and Rogers (1985). Recently work on these lines has been attempted for the Tambraparni and the Periyar–Vaigai basins in Tamil Nadu.

3. Lack of data is a serious but not insuperable obstacle. The principal problem is that the irrigation establishment has shown little interest in using these models for operational purposes and therefore made no effort to improve the database.

4. These studies covered, among others, the Bhakra Nangal, Cauvery Mettur, Hirakud, and Saradha canals. There are only two instances where the project command area has been resurveyed: the Godhavari–Pravara canal and Hirakud both by the Gokhale Institute of Politics and Economics, Pune.

5. An alternative technical solution envisaging, (a) a lower height for the Sardar Sarovar dam without affecting the quantum of water harnessed, and (b) a different pattern of spatial distribution of benefits, has been prepared by non-government experts (Paranjape et al., 1995). The problems of rehabilitation have received widespread attention due to the Narmada Bochao Andolan and other NGOs. The issues involved, the agitations, and their impact on government policy are reviewed in Dreze et al. (1996).

6. See Ch. 2; also Sengupta (1980); and Vaidyanathan and Janakarajan (1989); Janakarajan (1996); Sivasubramaniam (1995) Rajagopal (1991). MIDS researchers have recently completed a study of some 80 selected tanks in Periyar Vaigai and Palar basins (Vaidyanathan (ed.), 1998).

7. The literature on user participation in management is extensive (Olson (1965) and Ostrom (1990) for example view it in the framework of collective action theory. Sengupta (1991); Johnson et al. (1995); Coward Jr. (ed.) (1980) deal with specific user managed as well as bureaucratic systems and with the case for (and experiences of) recent attempts at transfering management to users. Advocates of community participation in irrigation management rely heavily on the theories of collective action without sufficient recognition of the fact that user communities from part of a larger, interconnected system. The required organizational structure is therefore complex and any satisfactory study of its working must comprehend the relation between its different segments and layers, the role of user communities in making and influencing decisions at various levels, and the principles governing inter-unit relations.

8. See for example the Johnson et al. (1995), and GOI, PC (1992). There are however significant differences with regard to details.

9. This point of view is strongly advocated among others, by the World Bank and IFPRI.

10. For a description of these experiments see Johnson et al. (eds), 1995.

11. This section is a somewhat edited version of the Foundation day lecture of the Society for the Promotion of Wasteland Development given by me in 1991. It was published in *Wasteland News* that year.

12. See also Bali (1988) for an informative and insightful, critique of the watershed development programmes of both government and non-government agencies. It also contains many excellent suggestions.

13. The ensuing observations on organizations are based on information available up to 1990. The need for departmental coordination and beneficiary participation have since been reiterated even more strongly. The latest to do so is the report of the Hanumantha Rao Committee on wasteland development. A more radical suggestion is to merge all the wasteland, watershed, and soil conservation programmes into a single, integrated watershed programme under a unified administrature agency, with the field level teams planning and implementing schemes in close collaboration with local communities (GOTN, 1997). There has however been little change in the situation on the ground.

14. These are reviewed in numerous publications including SPWD (1989), Chopra et al. (1988); S. Joshi (1988); B. K. Joshi (1988); Fernandez (1989); Datye et al. (1998).

15. Use of remote sensing for watershed and wasteland development is spreading. For instance, the Institute of Remote sensing in Anna University,

Chennai has prepared land use, soil and land capability maps for selected blocks and watersheds, as well as the specific physical structures, with approximate location, for effective soil and moisture conservation in the selected areas.

16. The techniques evolved in this experiment have now been widely adopted for local area planning by elected Panchayat Raj bodies. In the process the need for technical expertise to help communities interpret the information and use it to formulate plans for future development has become apparent. These problems are now engaging the attention of the Kerala Sastra Sahitya Parishad, which has played a pioneering role in this programme, and the Kerala Government which is vigorously pursuing this effort.

17. For interesting, stimulating discussions of innovations in 'appropriate' technology to augment biomass through watershed development and using it as a source of energy and as a nutrient, as well as the basis for value adding activities see Datye and Paranjape (1998) and Datye (1997).

18. These deficiencies are shared by most evaluation reports, including Dhruvanarayana et al. (1990); B.K. Joshi (1988); Katar Singh (1988); Deshpande and Reddy (1991), Ninan and Lakshmikanthamma (1994); and Chopra and Rao (1997), and several others.

Some use comparisons of treated with untreated watersheds; some compare beneficiary with non-beneficiary farmers; only a few do a before–after comparisons. This may be the best that can be done in the absence of diachronic surveys of project areas, but is clearly far from satisfactory.

19. The organizational problems have been examined recently by the Govt. of Karnataka, the Hanumantha Rao Committee appointed by GOI and the Committee on Wasteland and Watershed programmes set up by Tamil Nadu.

20. While most studies emphasize beneficiary participation, the nature and scope of the problems involved are not considered in depth. They tend to focus more on the problem of inter-departmental coordination, rather than those of achieving negotiated consensus in the user community and local institutions to enforce them.

21. This section is an expanded and substantially revised version of my lecture at a national workshop on water management, India's Groundwater Challenge, held in Ahmedabad in 1993. It has been published in the *Indian Journal of Agricultural Economics*, 1995.

22. The location and observation of the characteristics of wells where water levels are being measured have not been published, nor the basis of the selection. There is some reason to believe that the numbers may be too small and that the sample not representative. In Tamil Nadu the observation wells are mostly drinking water well located in or near village settlements. There is reason to doubt whether they reflect the groundwater conditions in irrigation wells of the area. These technical issues need to be critically examined in order to ensure reliable, timely, and representative observations.

23. There aspects been inadequately researched. For a recent attempt see Janakarajan and Vaidyanathan (1998).

24. Tushaar Shah (1997) reports that farmers in north Gujarat have taken to forming companies to encourage shared tubewells. The companies are said to be numerous and working well. This is in line with the expected rational response. This phenomen has not however been reported from any other part of the country. Experiments with community wells under state auspices have not been particularly successful.

25. The crucial role of policy for pricing energy used for pumping to create a strong disincentive against over-exploitation is persuasively argued by Shah (1997); see also Moench et al. (eds) (1995).

4

ORGANIZATION AND MANAGEMENT OF WATER CONTROL IN CHINA

INTRODUCTION

The study of the organization and management of water control works in China is interesting for several reasons: in terms of the area covered, China's water control system is the largest in the world and comprises works of diverse nature, size, and complexity. The existence of wide variations in all these respects and in agro-climatic conditions between different parts of China, and of the contrast between China and other countries (especially in Asia) provide scope for exploring, in a comparative perspective, the role of agro-climatic conditions, techno-economic factors and the organization of agricultural production in shaping the form and the working of water control institutions. China has also a long and virtually unbroken history of constructing and managing such works. This history, dating back over 2500 years, is better recorded than in most other parts of the world and appears to provide a unique opportunity to examine the dynamics of water control institutions. Of special interest, in this context, is an understanding of the impact of the post-1949 reorganization of agriculture (on the basis of communal ownership and management of land) on the structure and functioning of water control institutions.

The historical literature on Chinese water control is extensive: Chinese scholars and administrators have left behind fairly detailed accounts of the technical and organizational aspects of water control at different points in time over the past two millennia. This material forms the basis of several recent studies of the history of irrigation and flood control in China, the most notable among these being Wittfogel (1957), Chi (1936), Perkins (1969), Needham (1971), Morita (1974), and Hamashima (1981).[1] Needham's focus was primarily on the evolution of the technology of water control and only incidentally on matters of organization. Wittfogel was more concerned with the latter and, in his early work sought to unravel the factors (with an accent on agro-climatic conditions) determining the importance of water control and the nature and scale of works undertaken. Subsequently, however, he shifted the emphasis to a generalization regarding the connection between high degree of dependence on water control and the despotic character of 'Oriental States'. This thesis has been challenged even in the context of China. Chi, on the other hand, sought to explore the connection between the shifting geographical locus of political power and its relation to the development and maintenance of water control in the different phases of Chinese history.

Institutional aspects of water control were the central concern of both Morita and Hamashima: Morita, besides assembling a considerable amount of factual material on water control organization, sought to further clarify the relation between the agro-climatic factors, the nature of the water control system, and the characteristics of institutions for managing them.[2] Hamashima[3] traces the evolution of water control institutions in one region and in that process throws light on the interactions between land tenure, the sharing of responsibility for maintenance, and the working of institutions. We also have some descriptive accounts of the structure and management of institutions in specific, local situations in pre-revolutionary China (Fei, 1939; Motonosuke, 1979; Myers, 1975).

Several recent works deal with the post-revolution period (Anony., 1985; Chao, 1970; Nishimura, 1971; Vermeer, 1977; Nickum, 1976, 1977, 1981; Greer, 1979). Some of them (like Chao) are concerned with assessing broad trends in the extension and improvement of water control, the policies concerning these activities, and their impact on production. The institutional aspects of the construction of projects are dealt with in considerable detail, notably by Nickum, Nishimura, and Vermeer. Information on the working of institutions in operating the projects, especially in the management of water allocation, is limited in scope and detail. Much of what is available has been compiled by Nickum based on field visits and official Chinese publications. His translation of original Chinese documents on the management of particular projects (Nickum (ed.), 1981) is especially valuable.

No systematic or comprehensive review of this literature is available.[4] Given the widely different preoccupations and perspectives of these studies, a general summary of the material may not be of much interest to those interested in the organization and management of water control. What we need is a systematic collation of the information and insights bearing specifically on these institutional aspects and to present them in an organized form. This is what this chapter seeks to do.

My concern is to map the way the construction, maintenance, and operation of water control systems have been organized in China, the factors that shape them, and the manner in which these organizations work from the viewpoint of agricultural production. Since the nature of water control organizations depends on the nature and scale of the physical facilities, I open with a general description of the characteristics of the Chinese water control system. These characteristics are in part conditioned by natural factors (climate, topography, etc.). The inter-regional variations in agro-climatic conditions, and in the characteristics of the water control system are then highlighted.

In reviewing the nature of organizations involved in water conservancy and its management, it is useful to distinguish between the construction of the works and their operation. The operational function includes routine repair and maintenance of the facilities to keep the system in good working order and the regulation of the use of the facilities (especially irrigation) by laying down the principles of allocation between beneficiaries, establishing organizations and procedures to enforce these principles, and to resolve conflicts between users. These activities are of course not independent of one another and are often dealt with through the same, or overlapping, organizations and personnel. The important thing however is that they are functionally quite distinct.

The manner in which these principal activities are managed is influenced not only by the nature and scale of facilities serving a given area, but is also closely related to the way production is organized and output shared. Changes in land tenure and agrarian structure have no doubt occurred throughout; and they must have influenced water control (see Hamashima, 1981). However, such changes as took place after 1949 are quite unprecedented in scope and scale. Before the Revolution, China, as most other Asian countries today, was marked by private ownership and operation of land; unequal distribution of available land both in terms of actual cultivation and even more so in terms of ownership; and cultivation in small holdings (largely by tenant farmers). After 1949 a radical land reform programme was followed by the organization of agricultural producer cooperatives that soon gave way to the commune system. Though the system has undergone changes, and is again in a flux, it has remained the central feature of agrarian system of China over the past twenty-five years.

The changes in agrarian organization had several important consequences: (1) The traditional centres of power in rural society (the landlords and the gentry) that used to play an important role in all aspects of rural life, including water control, were replaced by new foci (namely the communes and the party cadres). (2) The ownership of land was communalized and principles for sharing the produce from land within villages underwent radical change. This led to more even distribution of incomes within village communities. (However, the inequalities arising from differences in per capita land and output across districts and even across villages within a district were not tackled as effectively). (3) The average size of the 'farm', viewed as a decision-making unit, was considerably increased. It was to be expected that these changes, together with the great emphasis on communal control of resources use in agriculture, would substantially modify the structure and working of water conservancy institutions. That is why it is necessary to draw a sharp distinction between the way various activities connected with water control were managed before and after the revolution.

WATER CONTROL IN CHINA: SOME SALIENT FEATURES

Historical Evolution

Irrigated agriculture was practiced in north China as early as the eighth century BC, possibly even earlier. In the early stages irrigation was probably on a very limited local scale. There are records of fairly large systems constructed between the sixth and the third centuries BC: One of them in North Anhwei province was reported to irrigate nearly 400,000 acres! These were however exceptional. According to Chi, under the feudal land tenure system then in vogue, there was little by way of 'surplus' or 'unattached' labour that could be mobilized for large-scale public works without interfering with the cultivation of the Lord's land. It is believed that conflicts over the distribution of the large increase in productivity due to the introduction of iron tools, ploughing, and animal manures around the fifth century BC led to the gradual disintegration of the feudal order, establishment of private property in land, and an increase in the volume of unattached labour. By the second century BC:

There were literally hundred of thousands of wandering labourers roaming the country and the system of private land ownership with a more or less fixed minimum tax burden obscured the danger of mobilization of the peasants for public works [Chi, 1963: 62].

This made large-scale mobilization of labour easier. Perhaps too,

The inefficient tools and low engineering skill of the period which preceded the introduction of iron tools prevented the construction of canals or tanks of any considerable size [ibid.].

Since then, as can be seen from the table prepared by Chi, the construction and improvement of these works has been a more or less continuous process. There were several contributory factors: an important one was the competition among rulers of different regions in developing irrigation and flood control works presumably because the resulting increase in production added to their resources and power. Water control works were sometimes used as a weapon in the struggle between regional rulers. The pressure of a growing population as well as food shortages in times of drought and flood were constant spurs to extend and improve water conservancy facilities. But irrigation and flood control were not always the only aims of water related public works. Thus in north China the need to transport grain tribute from the countryside to the capital and the requirements of defence were important considerations underlying the construction of several major canals. It is noteworthy, however, that even canals constructed for defence or transportation were used for irrigation by farmers along their course.

Progress, however, was neither sustained nor smooth: there were periods of vigorous expansion and periods of neglect and decay that coincided with the strength and vigour of the government at any given time. Water control was often a source, and on occasion an instrument, of political conflict. The geographical focus of these works shifted periodically in response to migratory movements and shifts in the balance of power between regions. Chi's thesis was that the regional power balance itself was significantly influenced by the differences between rulers in the seriousness of their interest in, and their ability to construct and manage, water control works.

Some Distinguishing Characteristics

At the turn of the twentieth century, China was estimated to have had close to 200,000 miles of canals, several thousands of miles of dykes and embankments, and reservoirs, of which there was an immense number and variety, covering some 13,000 sq. miles (King, 1912: 101–4). There was not much new construction during the first half of the century and the maintenance of existing facilities suffered from neglect. The military strife during the Sino-Japanese war and the subsequent civil war caused extensive damage to the water control systems. Since the establishment of the People's Republic, there has been an unprecedented, and sustained, effort at rehabilitating and improving the old systems and in construction of new ones.

Between 1950 and 1973, nearly 2,000 large and medium reservoirs storing over 10 m. m^3 of water (in comparison to barely 20 existing at the

TABLE 1

The Historical Development and Geographical Distribution of Water Control Projects in China

DYN PER / PRO	Spring & Autumn (722–481 BC)	Warring States (481–255 BC)	CH"in (255–286 BC)	Han (286 BC–AD 221)	Tsin (265–428)	Three Kingdoms (221–265)	Southern & Northern Dynasties (428–589)	Sui (589–618)	T'ang (618–987)	Five Dynasties (987–968)	Northern Sung (987–968)	Southern Sung (1127–1288)	Sung (Misc. Data)	Sung (Total for whole Dynastic Period)	Kin (1115–1268)	Yuan (1288–1368)	Ming (1368–1644)	Ch'ing (1644–1912)	Total items for each province	Date of Publication of Gazetteers
She	—	—	1	18	4	2	—	9	32	4	12	4	4	20	4	12	48	38	208	1735
Hon	—	3	—	19	4	10	—	4	—	—	7	—	4	11	2	4	24	843*	947	1767
Sha	—	—	—	4	—	—	—	3	32	—	25	—	—	25	14	29*	97*	156*	389	1734
Chi	—	—	—	5	2	—	3	—	24	—	28	—	—	28	4	11	223	542	886	1884
Kan	—	—	—	—	2	—	—	—	4	—	2	—	—	—	—	2	19	19	58	1736
Sze	—	—	—	2	—	—	—	—	15	—	—	—	5	5	—	—	5	19	53	1815
Kia	—	2	—	—	2	3	8	—	18	—	43	74	—	117	—	28	234	62	595	1736
Anh	—	—	—	—	3	3	4	—	12	—	7	9	—	16	—	30	30	41	127	1877
Che	—	—	—	—	—	2	4	2	44	—	88	185	31	302	—	87	480	175	1486	1736
Kia	—	3	—	—	—	—	2	—	20	—	18	36	2	56	2	13	287	222	658	1732
Fuk	—	—	—	—	—	—	—	4	29	—	45	63	294	402	—	24	212	219	1294	1754
Kwa	3	—	—	—	—	—	—	—	—	4	16	24	4	44	6	35	165	382	536	1822
Kup	—	—	—	—	—	—	—	—	—	—	—	—	—	—	—	6	143	523	728	1921
Hun	—	—	—	—	—	—	—	2	4	4	5	14	2	21	—	3	51	183	289	1885
Yun	—	—	—	—	—	—	—	—	4	2	—	—	—	—	—	7	118	292	412	1736
Tot for Dyn	6	8	1	56	16	24	20	27	254	13	290	543	363	1116	24	309	2270	3234		

Note: Reproduced from Chi, 1963: 36. These estimates have been critically reviewed and modified by Perkins (1969). The essential point here, namely the long history, and shifting regional focus, remains. DYN PER refers to Dynastic Period and PRO refers to Provinces.

time of liberation) and 'tens of thousands' of smaller reservoirs (storing 100,000 m^3 to 10 m. m^3) were constructed; 130,000 km of dykes were constructed or reinforced; nearly 100 large canals for diverting floods and draining water logged areas were dug; electrical and mechanical pumps with a capacity of 30 m. hp installed for irrigation and drainage; and over 1.3 m. wells sunk in north China to bring more land under irrigation (Chen, 1975). Large-scale investments in constructing and improving water control system have been undertaken since and are continuing. The total irrigable acreage in 1979 is placed at some 40 m. ha. in comparison to an estimated 12 m. ha. in 1949, and 23.5 m. ha. in 1954 (Kojima, 1982: 401).[5]

'Water Control' comprehends human intervention to regulate soil moisture in any one of three ways: irrigation (which augments the supply of water from rainfall and manipulates its timing and volume in accordance with the crop water needs); flood control (which consists of devices to prevent seasonal floods from inundating crop land); and drainage (which is intended to remove chronic accumulation of water in areas otherwise suitable for cropping). Flood control works comprising an elaborate system of dykes, embankments, and retention basins are to be found along all the major rivers, but particularly along the lower reaches of the Yellow, Hwai, and Yangtze rivers. Indeed some of the earliest water control works in north China were designed to regulate flood water of the Yellow river. The expansion of cultivation in some areas, notably near the estuary of the Yangtze, was made possible by drainage projects. In all areas of course irrigation is important.

The Chinese irrigation system is dominated by relatively small-scale, surface water based projects: in 1956, only 8 per cent of the irrigated area (34 m. ha.) was served by 'large gravity canals'. Gravity irrigation from 'small ditches and aqueducts' accounted for nearly a third of the total while nearly 40 per cent was served by 'farm ponds and weirs'. Groundwater irrigation, mostly confined to north China, accounts for the remaining 15 per cent of irrigated area (Table 2).

Similar data on area irrigated by different sources are not available for any subsequent year. However, the nature of the facilities constructed over the past three decades suggests that this pattern has not changed significantly: though the number of large reservoirs (with capacities of over 100 m. m^3) has risen from a mere 20 in 1950 to 308 in 1979, most of them appear to have been constructed primarily for flood protection and power generation rather than for irrigation. On the other hand, there are some 2100 reservoirs of medium size (10 m. to 100 m. m^3) and 82,000 small storages (0.1 to 10 m. m^3) whose the combined storage capacity is

TABLE 2

Area Served by Different Types of Irrigation Sources in China: 1956

Type of irrigation	(10^6 mou)
1. Gravity: large canals	43.2
2. Gravity: small ditches and aqueducts	182.0
3. Farm ponds and ditches	216.5
4. Pumping with electric or mechanical power	11.9
5. Well and other subterranean sources	86.4
6. Total	540.0

Source: Kang Chao (1970: 124).

probably much larger than that of the large reservoirs (Kojima, 1982: 401).[5] In addition, there are an estimated 6 m. farm ponds with a capacity of less than 100,000 m^3 (UN ESCAP, 1979). Besides increasing the number of ordinary wells and energizing them, some 1.6 m. power-operated pump-sets serving an estimated 10 m. ha. had been constructed up to 1976 (Vermeer, 1977: 202). There can be little doubt that the Chinese irrigation system continues to be dominated by relatively small-scale storages with groundwater playing a significant role in north China.

Another distinctive feature of the Chinese water control system is the widespread use of water lifting devices even for surface irrigation and drainage. Manually operated water wheels were common in the southern part of the country[6]; animals were used for lifting groundwater mostly in the north. According to one estimate (Dawson, 1966: 162) there were, in 1956, some 1.5 m. water wheels. This technique, which is both arduous and time consuming, has been rapidly replaced by mechanical pumps. In 1979, pumps for agricultural use (which would include lifting both surface and groundwater, and lifting for irrigation as well as drainage) were estimated at 71 m. hp. and nearly 60 per cent of the irrigated area was served by them (PRC, 1982: 122, 127).[7]

Regional Variation [8]

The necessity for different types of water control (namely, irrigation, flood control, and drainage), the technical characteristics of these works, and their relative importance is known to vary widely between different regions.

We have some quantitative information to show the extent of variation in irrigation in relation to cultivated area across provinces (Table 3 and

TABLE 3

Estimates of Cultivated and Irrigated Areas in Chinese Provinces 1957 and 1975

Province	1957		1975	
	Cultivated area	Irrigated area	Cultivated area	Irrigated area (10^6 *mou*)
North East				
Heilungkiang	109	4	109	14
Kirin	71	5	90	18
Liaoning	71	7	71	23
North West				
Sinkiang	30	26	44	31
Tsinghai	7	2	6	2
Inner Mangolia	83	12	83	10
Kansu	59	18	13	10
Shensi	67	10	75	18
North				
Shansi	68	11	63	10
Hopei	135	27	110	55
Peking	na	na	8	5
Tienstin	na	na	1	1
Shantung	140	37	115	53
Honan	135	43	110	48
East				
Kiangsu	93	38	75	51
Anhwei	88	24	70	40
Chekiang	34	25	29	23
Shanghai	na	na	5	5
Central				
Hupeh	64	28	60	39
Hunan	57	43	60	39
Kiangsi	42	29	43	(20)
South East				
Fukian	22	15	22	(10)
Kwantung	57	21	52	42
Kwangsi	38	25	42	22
South West				
Kweichow	30	8	31	12
Yunnan	43	12	41	15
Szechwan	115	55	116	40
China	1661	535	1547	656

Note: 15 *mou* = 1 ha.
Source: 1957 Chao (1970: Appendices 5 and 9).
1975 Vermeer (1977: 188–9).

Fig. 1), but scarcely any on the sources of irrigation or the nature and extent of other forms of water control, and can therefore only piece together a broad, qualitative, and impressionistic picture.

The nature and importance of water control is to a significant degree determined by agro-climatic conditions (rainfall, temperature, topography, soil, crops grown, and the like). China presents a great diversity of climatic and topographic conditions. Thus we find 'a hot desert climate in the Tarim basin, cold desert in Tibet, temperate continental climate in the north-east, the unique Szechwanese climate of the Red basin, and the tropical and sub-tropical in Kwantung' (Traeger, 1980: 14). Broadly speaking, the average temperatures (Figs, 2 and 3), average annual rainfall (Fig. 4) and its variability (Fig. 5) as well as the length of growing period

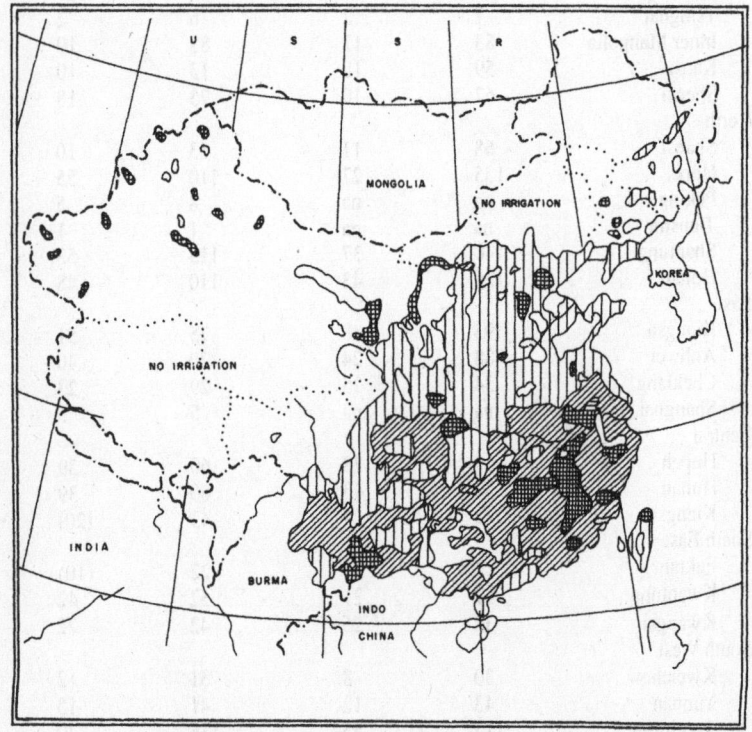

Fig. 1: Regional Distribution of Irrigation in China

Source: Cressy, George B., *Land of the 500 Million,* McGraw-Hill Company, Inc., New York, 1955, p. 91.

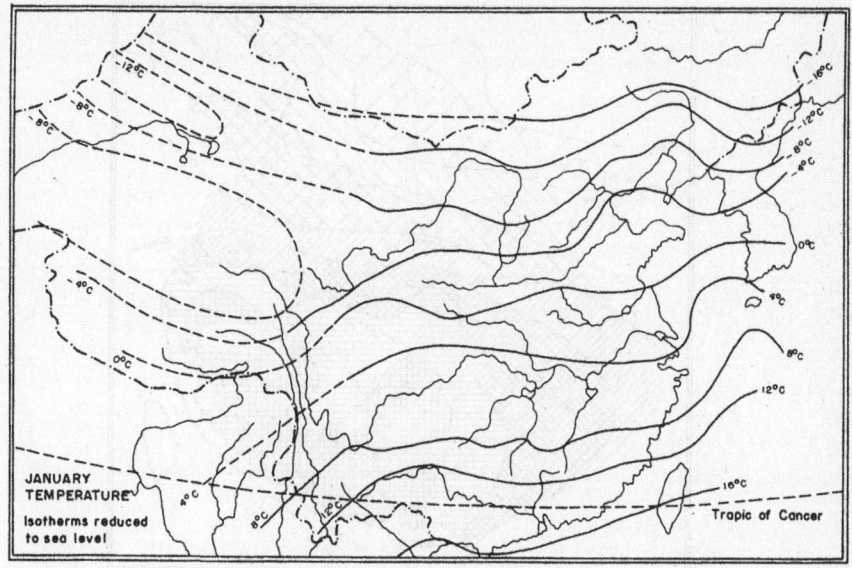

Fig. 2: Temperature: January Isotherms

Source: T.R. Traegers, *China—A Geographical Survey 1980,* p. 21.

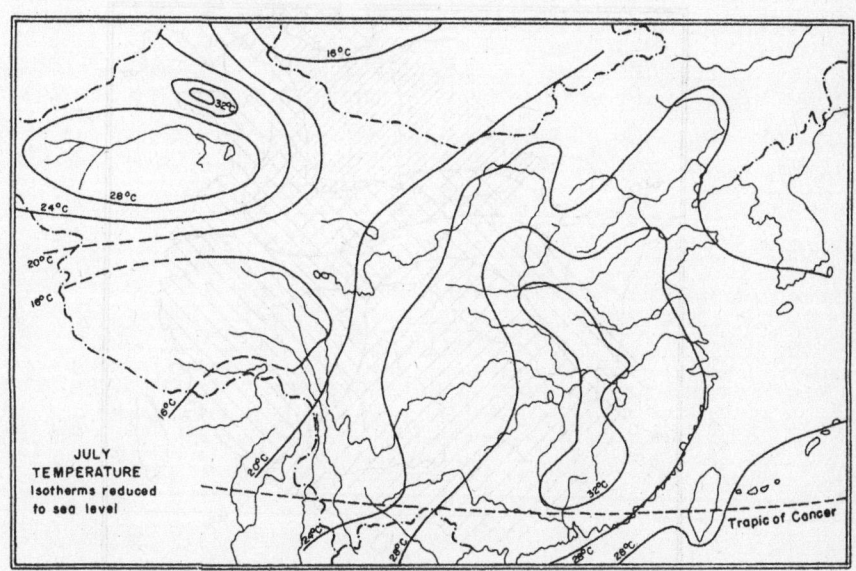

Fig. 3: Temperature: July Isotherms

Source: T.R. Traegers, *China—A Geographical Survey 1980,* p. 21.

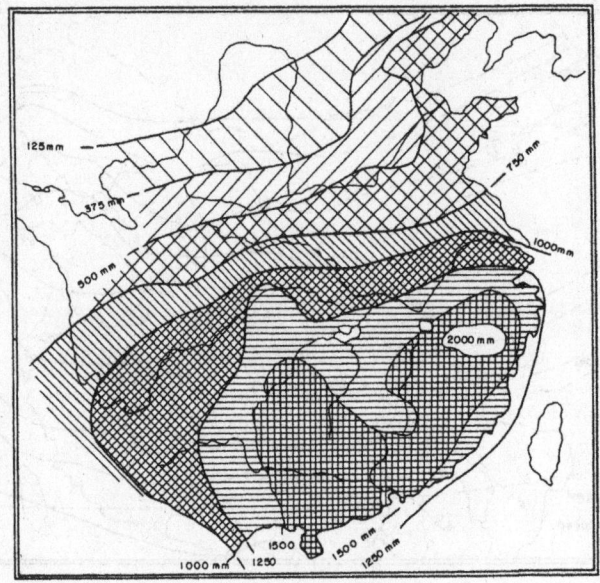

Fig. 4: Mean Annual Rainfall
Source: T.R. Traegers, *China—A Geographical Survey 1980,* pp. 24–14.

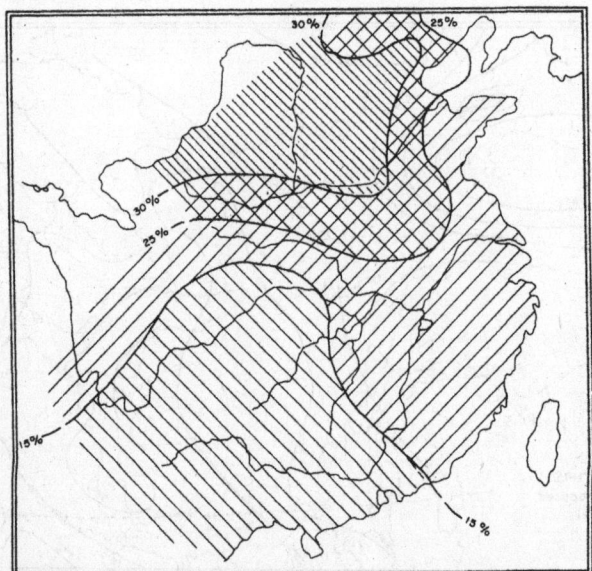

Fig. 5: Percentage Variability of Annual Precipitation
Source: T.R. Traegers, *China—A Geographical Survey 1980,* pp. 22–4.

(Fig. 6) all tend to fall as one moves from the south and south-east towards the north and west.

The mean annual rainfall in the north and north-east is relatively low (less than 800 mm) concentrated in short rainy season and also considerably more variable than in the central and southern parts of the country. The average annual rainfall in the Yangtze basin is between 1000 and 1250 mm while in the south and south-east the average exceeds 1500 mm. In the latter regions rainfall in also more evenly spread over the year and the variability in total rainfall between years is less than 15 per cent.

Togographic variations are also great: the terrain of much of the country is mountainous and rugged. Barely 10 per cent of the total area is cultivated, most of it concentrated in the valleys and estuaries of the major river systems, namely the Yellow river, the Hwai river, the Yangtze, and the Pearl river. All these rivers have their origins in the Tibetan plateau and the mountainous tracts of the central China and flow east or south, eventually draining into the China sea. The vast alluvial plains in their lower reaches and estuaries are among the most intensively cultivated and densely populated tracts of China. On the basis of climate and

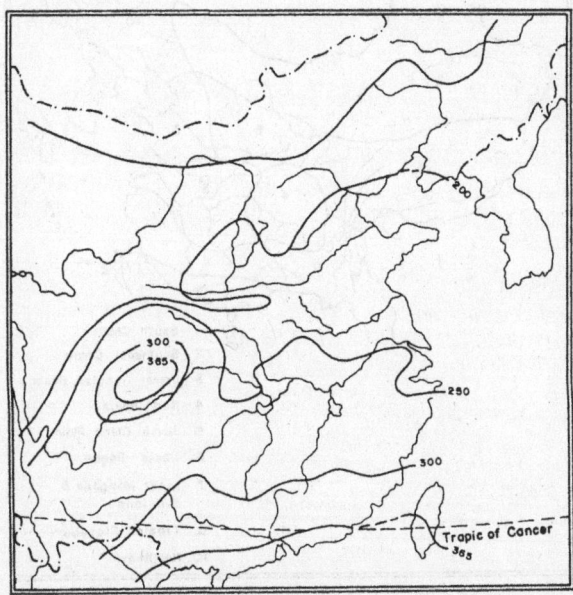

Fig. 6: Number of Frostless Days Per Year in the Lawlands
Source: T.R. Traegers, *China—A Geographical Survey 1980,* pp. 22–4.

topography, the country has been classified into several agro-climatic regions (Fig. 7).

The Tibetan plateau is far too cold and mountainous to sustain agriculture. The north-west, comprising Inner Mongolia and Sinkiang, is marked by a combination of severe winters and low rainfall which again limits the scope for arable farming. The north-east (roughly the old Manchuria) is a vast undulating plain but low rainfall (less than 750 mm) and severe winters (less than 200 days in a year are frost free on the average) limit agriculture to a single crop except in very limited areas where irrigation is available. Corn, millets, and soya beans are the principal crops of this tract. A notable development in recent times has been the rapid expansion of irrigated paddy cultivation in this region; the extent of area under paddy, however, seems to be small in relation to total arable land.

The Yellow river basin is also characterized by severe winters and a relatively short growing season. The average annual rainfall is less than 800 mm with an inter-year variability of 25–30 per cent. The rainy season

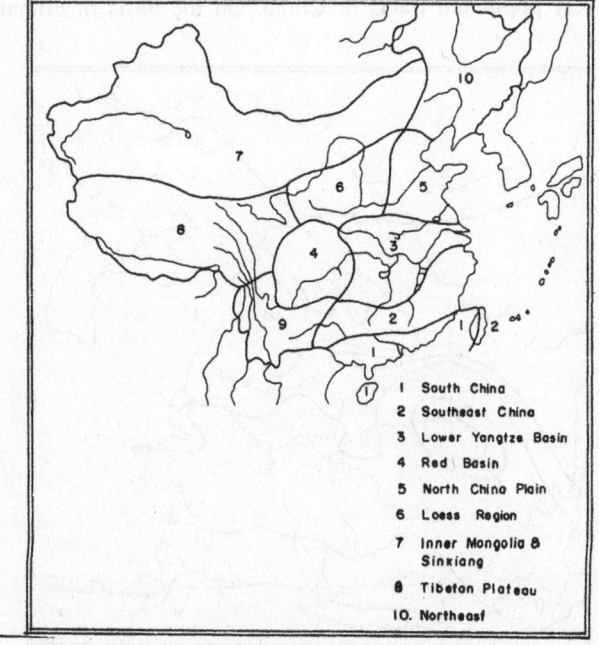

1	South China
2	Southeast China
3	Lower Yangtze Basin
4	Red Basin
5	North China Plain
6	Loess Region
7	Inner Mongolia & Sinkiang
8	Tibetan Plateau
10.	Northeast

Fig. 7: Climatic Regions

Source: T.R. Traegers, *China—A Geographical Survey 1980*, pp. 22–4.

is also much shorter than in the south. The basin, which was the centre of the early Chinese civilization, can be divided into two distinct parts: the upper portion, known as the 'loess region' and the north China plain. The former has fertile soils that need irrigation; and with irrigation the productivity of the land is high. Indeed, in this tract 'the valleys of the Wei and the Fen rivers have been irrigated from early historical times and formed the earliest granary of classical China' (Traeger, 1980: 215).

In the north China plain, by contrast, the problem has been and remains, basically one of controlling the enormous flooding of the river, regulating its wayward course, and capturing the silt in the river water to augment the fertility of the soil. Attempts to grapple with these problems again have a long history that has become legendary. From early times, the river was sought to be controlled by a system of embankments and dykes the scale of which was so large as to necessitate the mobilization of massive amount of human labour and organizing its works under the auspices of the state. However, because of the high rate of silt deposition[9] the embankments had to be constantly raised. Even this was not effective as the bed of the river steadily rose above the level of the cultivated fields. Breaches in the flood embankments were frequent and the problem was compounded by the waywardness of the river course. The construction of a series of reservoirs since the Revolution is claimed to have substantially improved the situation. These reservoirs are however silting up much more quickly than expected.

About two-fifths of the arable land in this tract is irrigated partly from surface water and partly from groundwater. Groundwater irrigation is much more important in the North China plains than in any other part of China. Most of the wells used for irrigation are in fact concentrated in this region. The post-revolutionary period has witnessed the construction of new wells, especially tube wells, on a large scale. The traditional technique of lifting water (relying on manual and animal power) has given place to mechanical pumps.

Wheat is the principal crop of this region followed by soya beans, tubers, millets, and corn. As in the north-east, rice is cultivated only on a small portion of the area but its importance is expanding as a result of the extension of irrigation facilities. Of greater significance perhaps is the reported success in raising, through a combination of changes in crop rotation and water management, the cropping intensity from three crops in two years to double cropping. On the other hand, inadequate drainage is reported to have aggravated the problem of waterlogging and salinity in lower reaches of the river.[10]

The upper reaches of the Yangtze basin are for the most part very mountainous and sparsely populated. The fertile province of Szechwan is an exception: surrounded by high mountains, its winters are mild and practically frost free. Rainfall (some 1000 mm) is well distributed and the area is fed by several rivers which are tributaries of the Yangtze. This region is also one of the centres of early Chinese civilization based on an ingenious 'system of terraced cultivation on the hillsides and irrigated agriculture in the valleys.' Nearly three-fourths of farmland is irrigated, and the use of water lifting appliances (water wheels in the past and, increasingly, energized pumps now) is widespread. An estimated two-thirds of the land is double cropped.[11] While rice is the principal crop, this tract is known for producing an exceptionally wide range of other crops, vegetables, and fruit.

The middle portion of the Yangtze basin, comprising the provinces of Hupeh, Hunan, and Kwangsi, is watered by the Yangtze and its major tributary (the Han river) both of which are liable to heavy summer floods. A system of dykes along the rivers and of linking the rivers to innumerable lakes along the banks is used to regulate the floods. Even so, summer floods submerge an extensive area. Rainfall (1200 mm per annum) is well distributed. The winter, though less cold than in the north, is severe enough to restrict the cropping season to 9–10 months. There is sufficient water in the ponds, lakes, and river channels to irrigate nearly two–thirds of the crop land. Much of the irrigation is by gravity from ponds located at the head of valleys. Often however water has to be lifted, and traditionally this was done by wooden paddle pumps operated by human labour. As elsewhere in China, mechanized pumps have replaced these traditional appliances to a considerable degree. Wheat, oilseeds, and beans are the major crops, and some two-thirds of the land is estimated to be double cropped.

Further east, and adjoining the coast, is a large plain built by the silt deposited by the river over several centuries and drained with the help of reclamation dykes. Powerful landed gentry and rich farmers used to extend their fields by building dykes and reclaim swampy land covered by lakes and rivers (Chi, 1963). Drainage is the major problem here, and for this purpose an intricate network of interconnected canals has been built. These canals also serve as sources of irrigation during the dry periods and as means of transport and communication. Part of the Grand canal connecting the Yangtze with the Yellow river is in this tract.

Though well-watered, year-round cropping is not possible because of winter frost. Nevertheless, a high proportion of land (estimated at two-thirds) is double cropped. Rice is the main summer crop followed by cotton, soya beans, and corn. These are grown in rotation with wheat, rapeseed, and beans in the winter. As in the north, the use of silt for fertilization is important but the technique is different. In the north flood waters are allowed to flow over the fields so that the fertile silt gets deposited in the cultivated areas. In the south the silt is captured in the beds of ditches, canals, and ponds. At the beginning of every season the silt deposits are removed and carried to the fields and spread. The quantities moved are enormous and involve a high labour input.

South China (comprising Fukien, Chekiang, Kwantung, and Kwangsi) has a tropical monsoon climate: winters are mild and frost free; summers hot and humid; and rainfall is relatively high, often exceeding 1,500 mm. Though over 70 per cent of the rainfall is concentrated between April and August, there is some precipitation almost every month. The mountainous and rugged topography of the region (especially in Fukien, Chekiang, and Kwangsi) restricts cultivation to a small portion of the area. However, Kwantung and the delta of the Pearl river sustain very intensive agriculture on the basis of an intricate network of irrigation canals drawn from the rivers. Most of the land is double cropped and a sizeable area is cropped throughout the year. Rice is the most important crop; but a great variety of other crops (including beans, sweet potatoes, vegetables, and tropical fruit) is grown. As in the Yangtze valley, efficient agriculture requires 'continuous draining of the fertile but swampy alluvial land and the maintenance of a complicated system of drainage and irrigation' (Chi, 1963: 12). I shall not discuss the regional variations in organization and management of water control. The necessary information is simply not available. Nevertheless the vast differences in the nature, extent, and role of water control between different parts of China should always be borne in mind.

ORGANIZATION OF CONSTRUCTION

The Pre-revolutionary Period

Historical accounts of the development of water conservancy works (Chi, 1963; Needham, 1971) tend to stress the role of the state in the construction and maintenance of large-scale projects. There were undoubtedly many such projects: the most notable being the dykes to control the floods along the Yellow, the Hwai, and the Yangtze rivers; the Grand canal; and several irrigation works capable of irrigating 100,000 acres and more. Besides directly contributing to the regulation of rivers and harnessing their waters for irrigation, it is quite likely that they created an environment in which small-scale dispersed works could safely be undertaken. It is also probable that during certain phases of the Chinese history and in some areas (especially north China), large-scale works under state auspices played a dominant role.

Projects of the size of the Grand canal and the embankments along the Yellow river naturally called for planning, mobilization of resources, as well as organization of the actual construction over a wide territory. Only the provincial or the imperial governments could undertake tasks of such magnitude. Since the techniques of construction were highly labour intensive, large-scale public works also required mass mobilization of labour. In the absence of a developed money economy and a free labour market, this meant 'forced labour assembled and disciplined by the authority of the state' (Chi, 1963: 123). Labour corvée for the construction of state-sponsored projects has been an important feature of water conservancy development from early times.

The scale and manner of mobilization is brought out by the following account of the construction of the Sui Grand canal: started in AD 609, an estimated 3.6 m. men were assembled to work on this project under the order of a Royal edict. Besides, each family was required to contribute one old or young man or woman to prepare meals for the workers. Altogether 5.4 m. people were employed on the project spread over several thousand *li*. They were supervised and disciplined by 5,000 soldiers, and after the completion of work in one sector, 2.5 m. labourers and 23,000 soldiers had been lost (Chi, 1963: ibid.). Another form of labour mobilization was through 'military colonization schemes' by which large teams of soldiers and workers were used to open new lands and cut canals in outlying areas. These examples are however inadequate to judge the role of the state or the nature and extent of the corvée labour system in public work construction (Chi, 1963: 99–100).

There were certainly phases of Chinese history (or, perhaps more correctly, one region or another during most of the Chinese history), in which the state and its bureaucracy played a major role in water conservancy construction. It needs however to be remembered that large-scale projects undertaken directly under the auspices of the provincial and imperial governments form but a small part of the total. Such projects were more conspicuous in the case of flood control along the major rivers involving the construction and maintenance of large embankments and dykes extending in some cases to over several hundred miles. In the case of irrigation, large-scale projects are relatively less important. I have already referred to the fact that less than ten per cent of the total area irrigated is accounted for by large gravity canals, and that the great bulk of irrigated area is fed by 'small ditches and aqueducts' and by 'farm ponds and reservoirs'. Perkins (1969: 171) has pointed out that the average size of project undertaken during the past 1000 years involved only a few miles of dykes and a few thousand acres of irrigation. He also cites data for two provinces to show that barely half the projects, for which information was available, were constructed by official agencies and the other half by private effort (ibid: 343).[12]

In more recent times there is ample indication that though there was a highly centralized bureaucracy, the planning and execution of works were in fact decentralized: most provincial governments had a Director General for river administration under whom there were a sizeable number of junior officials. Immediately below him was the River Intendant (of whom there was usually one in each province) and then there were a large number of magistrates of varying ranks specifically assigned to river administration work. The government also employed 'sluice keepers' who seem to have been responsible to the magistrates in charge of general administration of districts rather than for the river administration. In principle

water conservation and water works of the principal rivers, like the Yellow river and the Yangtze river, were under the official in charge of the river administration and were financed by government funds. But river tributaries, reservoirs and dams which affected merely irrigation of local farms were left to the local officials and the people. [Chu, 1962: 155.]

As a rule any large-scale repair or construction project (at whatever level) had either to be approved by a superior authority, or otherwise financed by the Magistrates (ibid: 180). As a matter of fact, however, the funds allocated by the state for such local water conservancy works were so limited that the magistracy had to, and did, depend to a large degree on local resources.

Several studies have cited evidence to show that, throughout local landlords and the gentry played a significant role in organizing the construction of innumerable local works independently of state projects or took advantage of the opportunities opened by the large state projects. Local institutions, dominated by these classes, also played a key role in maintaining and operating the local facilities. They were also often entrusted with the task of mobilizing corvée labour for state projects that had formerly been the responsibility of the magistrates appointed by the Imperial government.

Irrigation water in much of central China and south China... were supplied from locally built reservoirs and ponds.... Construction of reservoirs was first accomplished generally with a region's own labour either privately organized or conscripted by a local magistrate. [Perkins, 1969: 171.]

There are a few reported instances of local irrigation works constructed wholly by local landowning families. Motonosuke (1979) mentions two such systems in Hainan: one (covering 205 ha.) constructed by the biggest landowning family of the area during the Ming period (fourteenth to seventeenth centuries) after obtaining government permission and another relatively large system (2,460 ha.) established by a local landlord 'several hundred years ago'.[13] Mortia (1973) refers to a system covering some 300 villages which was set up in the eleventh century by a consortium of 14 landowning families.[14] Possibly the more usual pattern was for the gentry and landowners to play a leading role in promoting water control projects, lobbying for government support (including financial support) for such projects, and mobilizing local resources.[15]

There is ample indication that a substantial part of the resources needed for water control works was raised locally and that the large landowners and gentry played a significant role in mobilizing these. The methods of doing so were quite varied:

There are examples of many projects which were handled by individual members of the gentry and which were of great importance in the everyday life of the people. In projects covering a larger area, a number of gentry would pool their resources and abilities to plan and carry through the work. Members of the upper gentry usually took the lead.

Often the provincial officials stepped in to direct or to assist in coordinating the work of the various districts concerned. But whether the projects were directed by officials or by gentry, the latter shouldered the main burden of execution. [Chang, 1955: 57.]

Gentry members often financed such projects with their own money or with funds collected from local inhabitants... At times, however, they promoted and directed the projects but succeeded in turning over the financial burden to the government. [ibid.].

On the basis of concrete examples cited by Chang (1955) and Hsiao (1960) it would further appear that, in the nineteenth century, the proportion of the cost of water control projects borne by the government varied a great deal and that the balance was mobilized in the form of contributions, in labour or materials, by the beneficiaries. In some cases landowners supplied provisions and tenants provided labour. Hsiao (1960: 285–6) cites instances where part of the cost was financed by levying a special assessment on the owners and tenants of the area benefited.

... the expenses incurred were met partly by assessing cash payments on the land protected by the dykes, partly by gentry contribution and partly by grants from the government.

On the basis of the above it would seem reasonable to infer that, (a) the imperial and provincial governments played the major role in planning and implementing relatively large projects, with a substantial part of the resources being raised in the form of corvée labour; (b) the local centres of power (the gentry and the landlords) played an important role in persuading the state to undertake projects and in mobilizing the local resources needed; and (c) a large part of investment in water control (consisting of purely local projects and local works to make use of large state projects) was in fact undertaken on local initiative and with local resources either directly by, or under the leadership of, the landlord/gentry class. However there is an inadequate basis to judge the relative importance of different forms of resource mobilization. Nor is it possible to get a clear picture of how important corvée was in relation to other forms of labour contribution; on what basis these contributions were levied and enforced; the extent to which the freely hired labour was used; and the manner in which it was organized and remunerated.

The Post-1950 Period[16]

While, in a formal sense, the division of responsibilities between different levels of government in the post-revolution phase appears to be similar to that prevalent in the pre-revolutionary period, there are basic differences between the two periods in the way in which projects were identified, planned, and executed. The most striking, and in many ways unique, feature of the post-revolution period include: the attempt at decentralized planning and construction of projects through the agricultural cooperatives to begin with, and subsequently through the communes; the emphasis on the beneficiaries contributing most, if not all, the cost; the attempt to carry through a massive programme of construction with highly labour intensive methods; and the techniques used for mass mobilization of rural manpower without generating an additional demand for food. In all these respects, however, significant changes have occurred over the past three decades with the Great Leap Forward period marking the watershed.

Formally, at the national level, the Ministry of Water Conservancy and Electric Power and the Ministry of Agriculture share the overall responsibility for water control works; The former's functions include the preparation of plans for the basins of large rivers in consultation with the Ministry of Agriculture in matters involving agricultural irrigation;

the design of large-scale engineering works for water control; and providing technical guidance to the local authorities who are responsible for planning for smaller river basins and designing 'less important' projects.

The Ministry of Water Conservancy has its own specialized bureaus in Peking. The separate Commissions for the integrated planning of the Yellow, the Hwai, the Hai, and the Yangtze basins are also under its overall control. Provincial departments of water conservancy are responsible for planning and executing works within the provincial jurisdiction; and they also serve as the spokesmen for the interests of their respective provinces with the national Ministry in matters involving more than one province. Disputes between provinces concerning matters related to water for agriculture are mediated by the Ministry of Agriculture (Donnithorne, 1969: 126–7; Nickum, 1979: 171–3). A more recent account of water conservancy planning describes the position as follows:

The planning, design, construction and operation of specific projects takes place at four different levels, depending on the size of the project. The general principle is that the responsibility for a project which affects two or more units is taken up by the unit of the next higher rank. A corollary is that each unit which benefits from a project contributes labour and investment in proportion to its share of the benefit. Thus the central government, through the project bureaus of the Ministry of Water Conservancy, takes the responsibility for major dams and power station projects and labour to supplement state investment funds is contributed by all provinces that will benefit. Provincial governments are responsible for irrigation projects that affect more than one county or municipality. County (or municipal) governments usually undertake the diversions or reservoirs which affect more than one commune.... At the local level commune or production brigade units plan, construct and operate numerous projects of all types. [Greer, 1979: 116–7.]

Immediately following the Revolution, roughly over the period 1949–56, the emphasis was on restoration of facilities that had been damaged or fallen into disrepair from neglect during the long period of conflict. During the First Five Year Plan a number of large-scale projects, involving the use of relatively mechanized construction techniques, were taken up, especially in the Yellow and the Hwai river basins. The government financed a relatively high proportion of the outlays on water conservancy but no precise data are available. There are however indications that even during this period the volume of local effort, undertaken largely by the agricultural cooperatives organized as part of the land reform and under the active leadership of the party cadres, was by no means small. Indeed, the bulk of the increase in irrigated area in the first plan came from small, local projects. Large-scale projects under the state budget added, between 1949 and 1956, a mere 1.2 m. ha. to irrigated area in comparison to 8.4 m. ha. from 'cooperative projects of medium and small scale' and 6.2 m. ha. from wells, water-wheels, and pumps (Dawson, 1966: 159).

The volume of earth-work done under state-led water conservancy works, which gives a rough measure of the labour mobilized for construction outside the community, rose from an estimated 420 m. m^3 in 1950 to 1,700 m. m^3 in 1956 (Vermeer, 1977: 46). The volume of work involved in all categories of projects (i.e. state and local projects) must have been several times larger. A 1957 survey showed that on an average an agricultural worker spent 2.4 days per year for state projects, mostly water conservancy. This was in addition to an estimated 9.4 days per worker spent on farmland capital construction (which includes levelling land, water and soil conservation) not on behalf of the state' (Vermeer, 1977: 278).

Beginning in 1957, there were several important changes of policy marked by a shift from, (a) the prevention of floods to full utilization of water resources for irrigation; (b) drainage of water to storing it; and (c) larger projects, financed by appropriations of the Central government to small projects undertaken by local authorities and individual agricultural cooperatives (Chao, 1970: 125–6). The shift in favour of storage and irrigation is reflected, according to Chao, in the emphasis on increasing storage capacity by building ponds and reservoirs as well as in the tendency to cut flood protection dykes to draw water for irrigation. Vermeer cites some evidence, albeit based on information relating to only a few projects, that upwards of 70 per cent of the cost of water conservancy projects after 1958 have been met by communes and associated institutions. This appears to be substantially higher than in the pre-Great Leap Forward phase (Vermeer, 1977: 260–1).[17]

Equally important was the shift in the organizational sphere: besides increasing the reliance on local contributions by way of mass labour mobilization, there was a strong emphasis on giving local institutions a major role in planning and designing projects and in organizing their construction as well as on the need for 'collectivism' in this effort. The working of the new approach, in the early phases, is illustrated by an official document describing the experience of selected projects (some 7–8 of them) in different parts of the country.

In almost every case the 'technical experts' raised doubts about the technical problems of design and construction, and cautioned against the dangers of taking up projects on the basis of hasty designs without proper technical investigations. But in every case the caution was rejected. While technical considerations were important, speedy development of water resources was even more so.

> ... the technicians often sticked to conventional and old methods of work. Under the pretext of scientific requirement, standard quality, regular practice, etc., they frequently caused unnecessary delays. [Anon, 1958.]

The account used the experience of the selected projects to make the point that most technical difficulties could be overcome because peasants learn quickly from experience and 'collective wisdom works wonders'. The intimate knowledge that peasants had of the conditions and problems in their areas was repeatedly stressed as an advantage for proper planning.

The principle of self-reliance was of course emphasized even more in finding the resources needed for project construction. Even in the case of large, state projects, the beneficiary units (communes, brigades, or teams as the case may be) were required to provide the labour necessary for construction. As for works undertaken at the commune level or below, the cost of planning and construction had to be wholly met by the beneficiaries from their own resources. Construction work was as a rule taken up during the slack season, the necessary labour being mobilized through the communes and their constituent units. Labourers assigned for project work were not paid any wages directly but they received work points in their team which was also responsible for taking care of their families. In this way a large volume of labour could be mobilized for capital construction, apparently without generating additional demand for food and consumer goods.[18]

Initially production teams were required to contribute labour only for projects that benefited them, and the extent of contribution was determined on the basis of their share of the project benefits. The team decided who should be assigned for construction and for how long. These principles were however changed during the Great Leap Forward period and interpreted more broadly to mean that teams should be willing to contribute their labour even in the construction of projects that did not benefit them directly.

In the past when water conservancy works were to be built labour, material and funds were demanded from the people in accordance with the amount of benefit they were to receive from the projected works. This principle holds good under normal conditions. But the Hsushui people wanted to rid their county for good, in one years time, of the big floods... and severe drought. A different principle was therefore required. Accordingly a plan was drawn up for water conservancy for the whole county. Labour power was put under unified control and work was done according to schedule. [Nickum (ed.), 1981: 12, 13.]

Many instances are cited where people were persuaded, through 'political education', to contribute a large amount of labour to projects that did not benefit them directly.

The culmination of this approach came in 1958 which was marked by a phenomenal increase in the pace of construction and in the scale of mobilization: by the end of 1957 it was reported that there were 1.76 m. projects for canals and ditches and over 8 m. water-storage projects under construction in the entire country. Their number is believed to have risen steeply in 1958 but no data are available. (Chao, 1970: 128.) However, according to official reports, during the first few months of 1958 over 100 m. peasants were working every day on water conservation projects (ibid.). One report in the Chinese press placed the volume of earthwork in 1958 at 58 bn. m^3 and the volume of labour used at 70 per cent of the labour force working for half the year or more on water conservancy projects (cited in Vermeer, 1977: 45). While the reliability of this and other estimates is open to doubt, there is no question that the mobilization was massive. Most of this was for local works, but in absolute terms the amount of work that people had to do outside their brigades and communes, sometimes at far away locations, was quite large. There was no change in the system of remuneration for such work.

The programme, however, soon ran into serious problems for several reasons. The scale of mobilization was so large that it cut into labour supply available for normal cultivation operations. Moreover, the system by which brigades/teams were to look after the families of workers away on construction projects and giving credit to such workers in the form of additional work points, did not work satisfactorily. Also, resistance to the rather wide interpretation of collective contributions of labour seems to have grown. Besides, the technical defects in projects undertaken during the Great Leap Forward period (in the form of defective designs, poor quality of work, failure to construct distribution and drainage channels, neglect of repair and maintenance) were widely reported. (For a detailed discussion of these points see Carin, 1963; Chao, 1970: 130–7; Nishimura, 1971; and Vermeer, 1977: Ch. IV.)

These problems, together with a succession of bad harvests, led to a sharp cutback in the scale of water conservancy construction after 1959. New construction was practically discontinued after 1960 and attention devoted largely to repair and improvement of existing irrigation facilities; Vermeer (1977: 60–1) estimates that the volume of earth and stonework

in 1962–3 was less than 200 m. m³ or about one-ninth the 1957 level.¹⁹
The officially reported irrigated area, after recording a near doubling in
1957–9 fell sharply until, by the mid-sixties, it was no higher than it had
been in 1957. This is sometimes used as an indication that much of the
work undertaken during the Great Leap Forward period proved to be
'infructuous'. The changes could be due in part to changes in definitions
and to better reporting.²⁰

The period 1961–5 was one of consolidation, with the emphasis on
improving the effectiveness of works already done by completing the
distribution network and providing proper drainage. The total volume of
activity recovered gradually but by 1965 it was no higher than the level
recorded in 1957. During the Cultural Revolution period mass mobilization
was revived and the pace of construction of water control projects appears
to have increased significantly. The emphasis throughout this phase
continued to be on small (commune/brigade level) projects with an accent
on land improvement.

After the Great Leap Forward phase, there were some important
changes in the basis of labour contribution and the organization of
construction. In contrast to the earlier phase,

Only those production teams that will benefit from [a project] contribute labour;
when other teams contribute labour they receive compensation. For large projects,
the commune and local governments have established all-year round farm land
construction forces.... Contribution of peasant labour to very large projects... is
well regulated and controlled by Provincial governments. [Vermeer, 1977: 109–
11.]

According to Nickum (1979: 173), most of the farmland capital
construction teams set up 'to handle larger and more technically demanding
projects at the local level'

have been established at the brigade level, but some are set up at the commune
and *xian* levels as well. Organizationally, the teams are based on the military and
structured similarly. The 28 million members of these teams are remunerated by
work points in their home production teams, with supplemental subsidies at the
commune and *xian* levels. In addition some farmland construction teams grow a
portion of their own food.

During the seventies Nickum (1978: 282) reports a growing
concentration on 'Integrated development of entire water systems' and

greater reliance on capital intensive projects (like tube wells and small hydel stations). The bulk of the labour input into water conservancy construction continued to be absorbed by such traditional activities as terracing, construction of earth/stone dykes and canals, and levelling of land. The scale of mobilization has risen rapidly after 1970. Nickum (1978: 280) estimates the total volume of earth and stonework in farmland and water conservancy construction to have risen from 3.5–4.5 bn. cu. m per annum during 1965–70 to 6 bn. in 1973–4, and a high 25 bn. cu. m in 1975–6. Ishikawa (1981) cites data for one province where the number of rural workers mobilized for construction was only 3 m. in 1970 but rose to 13 m. (roughly half the rural labour force) after 1975.

The assigned works were so large that they once again began to cut into the time available for production team members to participate in normal cultivation and to carry on their subsidiary occupations (that included tending their private plots). In consequence there was, according to Ishikawa (ibid.), a revival of 'coercive assignment' of tasks for irrigation and farmland construction work after 1973; the allocation of the burden of mobilization among production teams was made without paying attention to the 'mutual benefit principle'; and the share of labour used for the relatively large projects under the district (or Hsien) auspices rose sharply after 1970 while that of projects undertaken by the communes/ brigades fell. In consequence, the difficulties of, and resistence to, mobilization of labour for construction is said to have again become acute. Then came the more basic shift in policy in 1979.

By separating the economic and the administrative functions at the commune level, the 1979 reforms weakened the party cadres' ability to influence the economic decisions of the brigades/production teams. At the same time, the principle of collective cultivation of land owned by a team and sharing of the common produce was diluted with the introduction of a system whereby the team was allowed to contract with individual members or a group of them to cultivate their land in return for specified minimum payments. Such 'contract farming' spread rapidly. The declared policy regarding construction and maintenance of water control works reasserts many of the elements of earlier policy, namely self-reliance, labour contribution by beneficiaries, and maintenance of specialized construction teams.[21] However, whether and how far this could be, and has been, sustained in the face of the loosening of political control and weakening of the principle of collective management of farming remains unclear.[22]

MAINTENANCE AND OPERATION OF WATER CONTROL SYSTEMS

The Organizational Structure

The Pre-revolutionary Period

There were hardly any systematic or comprehensive accounts of the way the maintenance and operation of water control works was organized in pre-revolutionary China. We can only reconstruct a general, and necessarily sketchy, picture from a variety of scattered sources. Caponera (1961), tracing the evolution of water laws in China, has shown that from ancient times, (a) the state's right to control the development and use of water was recognized; (b) a part of the bureaucracy was specifically assigned to the task of periodically inspecting the condition of water conservancy works, getting the necessary repairs done, and ensuring equitable distribution of the benefits; (c) local institutions connected with water control were subject to supervision by the state bureaucracy; and (d) there were fairly clear general principles governing the rights of individual users *inter se*, and vis-à-vis the state. As early as the AD eighth century, possibly even earlier, many of the matters had been formally codified. It would also appear that the principal features mentioned above have continued largely intact down to recent times.

Some idea of the size and structure of the state bureaucracy for water conservancy management is available for the nineteenth century (see Ch'u, 1962; Chi, 1963). In each province there were a number of assistance magistrates/sub-prefects and registrars in charge of river administration under the supervision of a River Intendant who was in turn responsible to the Director General of River Conservation. There were also sluice keepers 'in places where water control was necessary'. A compilation made in 1899 (ibid.: 11; 208 n. 70) suggests that at the district (*hsien*) level the number of assistant magistrates assigned to river administration (which seems to include both construction and maintenance) was nearly as many as the number appointed for general administration. The magistrates had formal responsibility 'to see that the rivers were kept properly dredged and dams kept in proper condition so that local people could have the benefit of irrigation' (ibid.: 155).

In reality, however, the effective reach of the bureaucracy in managing maintenance and operation must have been quite limited because, in relation to the area covered by the Chinese water control system, the number of officials was quite small; moreover, as we shall see presently, they did not get enough resources to carry out their tasks. It seems more likely that the state bureaucracy confined its attention to managing large projects and key facilities (like reservoirs, large dykes, major canals and sluices) set up by the state and affecting extensive areas, leaving the rest (which would cover the smaller canals, distributaries, and dykes, as well as systems constructed by local effort) to be managed by local institutions.

with a few notable exceptions the historical Chinese state concerned itself primarily with manipulating water for purposes other than irrigation especially flood control and navigation... on farm application and removal of water was left in the hands of local villages and inter village organizations. [Nickum (ed.), 1981: 3–4.]

What exactly the relative roles of the two were, whether, and in what manner, they interacted with each other are questions about which little seems to be known.

The local organizations, according to Nickum (ibid.),

varied in form from place to place and from time to time. Sometimes they were built and run by families separate from land holdings. At other times, especially in recent centuries, they were operated by landlords or other farmers associations, usually with full time managers.

We have accounts of only a few such local-level irrigation organizations.

One of these relates to a multi-village irrigation system in north China (Myers, 1975: 198–205).[23] This system, 'constructed in the sixteenth and seventeenth centuries by peasants of the area mobilized by officials and literati' consisted of a series of water gates and sluices along channels leading away from the main rivers flowing through the region. The management of the system (including maintenance and water allocation) comprised of two managers, one of whom was designated as chief supervisor, assisted by a number of 'subordinates' (roughly at the rate of one for every 20 households). The managers, who are said to have represented all the user villages, were 'middle-aged and elderly farmers who were very experienced in water management.' They were also mostly from the relatively large landowning families and many of them seem to have held office on a hereiditary basis. The number of managers depended on the size of the system; where it was large, the chief supervisor and his assistant were selected annually by the managers from among themselves. Each year the subordinates selected their successors 'on the basis of demonstrated leadership and skill in supervising other families, mediating disputes and loyalty.' In addition, there was an independent inspector, apparently a permanent appointment, to inspect the water gates and sluices and report on the repairs necessary.

Montonosuke (1979) briefly describes two small systems in Hainan.

Both were old systems set up by local landlords. At the time of the study (1942), the supervision of the system in one case was under the control of the largest landowning family, the chief and the vice-chief of the supervising committee being members of the family. Four other members could in principle be from other families but were appointed by the chief supervisor who was also responsible for mobilizing the pesants for maintenance. The other system was larger (2,460 ha.) and again traditionally controlled by the landlord family. Just prior to the Sino-Japanese War, the supervisor of the system was appointed by the county government which by then had a section for water control administration and mediated in disputes between villages and groups. However, the supervisor belonged to the landlord family. The system was divided into 6 parts each of which covered 10 distributaries. Every part was managed by a committee, whose heads were traditionally drawn from the landlord family, consisting of 10 members each responsible for supervising one distributary.

Evidently there existed a considerable variety of patterns, including centralized management by the state bureaucracy, wholly private 'control' and management, community systems operated by elected representatives of user members, and many intermediate types. The size of projects was no doubt an important determinant of the type of management: for instance, the state it appears was likely to play a more important and active role in relatively large systems. The larger the systems the greater the need for a 'bureaucracy', though this bureaucracy need not necessarily be constituted by officials of the state. There were however other factors besides size. For instance it has been said that 'traditional forms of village-level water management were all grounded in private or clan ownership of water rights on land' (Nickum (ed.), 1981: 13). As an official Chinese publication in 1965 asserted, before the revolution

In management organization, water rights were entirely controlled by the bureaucrats, the landlords and local despots and the distribution of water was very unfair and unreasonable. [Ibid.: 54.]

Several other factors (including the nature of the works involved, land tenure, and the broader framework of political and administrative institutions) must have also played a role in shaping both the formal structure of the organizations and, more important, the way they actually functioned.

Changes Since 1949

Over the past three decades there have been far-reaching changes in the organizational structure for maintenance and operation, and the way they are managed. To begin with, all forms of private (i.e. individual landowners, clans, etc.) rights over water were abolished as part of the land reform. The right to develop water resources and regulate them came to be vested with the state or the cooperatives (and later the communes).[24] The organizational arrangements at the user level however took some time to develop. Soon after the revolution, attempts to rectify the derelict state of water control management appear to have concentrated on expanding and strengthening the bureaucracy of the water conservancy departments at all levels of government. It was however soon clear that this was not enough and that there were serious weaknesses in the lower levels of management. While land reform had dismantled the traditional institutional framework in the villages, nothing had yet replaced it, and effective management was not possible without the bureaucracy working in close collaboration with, and thorough, the beneficiaries.

The fifties and sixties saw the rationalization of the organizational structure as well as several important innovations for coordinated operation of different parts of each system under a management that would be professional, unified and, at the same time, provide for effective participation of the beneficiaries' representatives at the various levels. The general approach to the organizational question is set out in two offcial publications: *Irrigation Management* (1965) and *Farm Water Conservancy and Soil Conservation* (1965). Excerpts from these publication are available in Nickum (ed.), 1981: Readings, 1, 2 and 11. We also have a description of the actual experience in building effective management systems and procedures for a few projects (ibid.: Reading 3).

The pattern of organization evolved during this period has several distinctive features. In the first place, the functions of these organizations are conceived in very broad terms:

The main tasks of project management are: to guarantee the completion and rational operation of projects, to extend the operational lifetime of the projects and to promote and expand the effectiveness of projects. [Nickum (ed.), 1981: 59–60.]

More concretely, this involves (a) 'the formulation and implementation of annual control and operational plans'; (b) maintenance of projects 'in good condition' to prevent them from getting damaged or minimize such damages; and, (c) improvement and rebuilding of projects.

Secondly, the organizational design seeks to provide a strong and unified management that combines professional management with participation by the broad masses of water users.

a strong management structure must be established at every level. The management structure should rely on the broad masses... set up various management rules and regulations and formulate plans for production and finance. It should be responsible for carrying out related party and government policies and technical measures and, using water scientifically, maintain project facilities. In order to facilitate management and make it relate closely to the masses served, it is necessary to rely on the poor and middle peasant masses benefited to build up a mass democratic management organization representing water using units in the irrigation districts. [Nickum (ed.), 1981: 61.]

The necessity of having a strong core of professional management personnel is recognized: There are indeed concrete guidelines on the size and composition of professional staff considered desirable for different categories of projects (ibid.: 65–6). At the same time, since the management of the water control system affects the production and livelihood of peasants and also involves a great deal of work, it is essential that professionals work with, and through, the masses: 'It will not do to rely merely on a few professional personnel to handle the matter in a cool aloof manner.' This requires, (a) mechanisms by which the interests and concerns of the users can be brought effectively into the process of making and implementing decisions; (b) arrangements for the professionals and users to work together; and (c) a clear delineation of functions, responsibilities, and sanctions.

The organization for managing projects is multi-tiered. The overall management of a project is, as a general rule, controlled by the 'lowest administrative unit whose boundaries encompass the benefited area.' The criteria for determining which level of the state or the collective sector, as the case may be, is to have responsibility appears to be fairly well-defined.

For state-managed irrigation districts which service areas within a single *hsien*, district, or province, the respective *hsien*, district, or province will set up a separate management organization. When the areas benefited cross *hsien*, district, or provincial boundaries, when the area served is within one region but the irrigation district is comparatively large and the relationships involved are comparatively important, then the next higher level sets up a structure responsible for management, or the various units served, under the direct leadership of the next higher level, organize a joint structure responsible for management. When the area benefited by an irrigation district which is managed by the commune sector crosses commune lines, the relevant communes organize a joint management structure subject to the leadership of the *xian*. When the area benefited is within one commune, the commune sets up a special structure to be responsible for management. When the area served crosses brigade lines, then under the leadership of the commune, the relevant brigades (*dui*) organize a joint management structure to carry out management. [Nickum (ed.), 1981: 64.]

The primary operational unit for management is in all cases the production team. Between the team and the body responsible for managing the system as a whole there are usually several layers of decision-making and implementation. In the case of projects wholly under the collective sector, the intermediate levels appear to be defined along territorial divisions, namely brigades and communes/municipalities. In larger projects they are demarcated sometimes on a territorial basis, sometimes on the basis of the physical component of the system (main canal, branch and sub-branch canals, pumps, etc.), and occasionally both.[25]

The management bodies of brigade and commune level projects are as a rule constituted from amongst their members. In some cases a part of the management committee/group is nominated by the higher level body (commune in the case of the brigade, the district in the case of communes) while others are elected.[26] In larger projects, management committees at the intermediate levels and at the top comprise representatives of user units (generally drawn from the persons responsible for water management at the team, brigade, or commune, as the case may be), professional staff under the control of the central management in charge at that particular level; and nominees of the Party Committee at that level.[27] The committees are the 'mass democratic organizations' through which the key problems in operating the system are tackled jointly by the users and the professionals, with the professionals, explaining to the users the technical aspects of problems and the ways of resolving them, and the users articulating their problems, experience, and interests. The expectation is that over a period time, this process will enhance the mutual understanding of, and sensitively to, each others' concerns, thereby facilitating consideration of issues in a much broader perspective.

The operating personnel consist of the professionals under the control of the central management and those appointed and controlled by the organs of the collective sector. Initially the lower level units did not have personnel with the technical knowledge and skills necessary for proper handling of water control tasks, such knowledge and skills largely being concentrated among professional staff. An essential part of the organizational reform has been to build a cadre of people in the production teams, brigades, and communes trained in the techniques of water management. Besides there are also specialized technicians (such as pump operators and mechanics) and personnel to perform 'watchmen' functions.[28] It is considered desirable for some members of the teams, brigades, and communes to specialize in water control management. Whether or not they should be exclusively assigned to water control work on a full time basis appears unresolved.[29] Also noteworthy is the fact that in the choice of personnel for water control, whether for managerial or technical functions, qualities such as possession of good ideology and good health, aptitude for a water control work, knowledge of local conditions are stressed.[30]

The professional management personnel, who appear to be drawn from officials of the concerned government departments (like the prefectural or Provincial Water Conservancy Bureaus), are of course expected to carry out the decisions and directions of the management Committee/ Conferences of the district as a whole. They however also have an important role in the process of making decisions to promote the objectives of unifed planning and coordinated implementation as well as of improving the effectiveness of the system in raising production in accordance with the larger directive of the Party and government. They can, if necessary, draw on the advice and expertise of the higher level government departments. Personnel assigned to implement water management tasks in the collective sector are expected to work in close collaboration with the professional management personnel at the particular level, but their responsibility is to the management committee at that level and not to the professional management. Strengthening of user level organizations and personnel is meant to relieve the professionals from routine tasks so that they may concentrate on more important aspects of management. In one case it is reported that as a result of this:

The number of professionals has been reduced, their contacts with the broad masses are much closer than in the past and their work penetrated much deeper [Nickum (ed.), 1981: 198].

The principal tasks of the 'mass democratic management organizations' involved in water control management (the district management committee, the main and branch canal committees, and the water conservancy committees/sections of the brigades) are:

convening meetings at fixed intervals to study and discuss relevant matters of importance such as project repairs, water utilization plans, canal system rebuilding, the levy of water fees, and formulation of relevant contractual systems. The matters discussed and decided upon by the conference of delegates and management committees are reported to higher level leadership departments for approval and then implemented. [Ibid.: 70.]

The requirement that lower level decisions have to be approved at the higher level before implementation is but one of many devices to ensure coordination of activities of different levels. In addition, the procedures provide for system of inter-level consultation and exchange of information so that the situation and concerns at a lower level are taken into account when formulating overall plans, and at the same time the lower levels are informed of the overall position of the project.[31] The practice in one project is described as follows:

Every year before spring irrigation and winter repairs, the Conference of representatives is convened to discuss problems relating to management. At the conference the opinions of the broad masses in the irrigation district are made known and the Party's Water Conservancy policies and the conditions in the irrigation district are communicated to the masses *so that everyone will be concerned about the overall situation and know that is going on.* [Emphasis added; Nickum (ed.), 1981: 197.]

The professional management personnel who are members of the intermediate level management bodies are expected to place the local problems in the context of the overall project situation. Periodic supervision and inspection of activities at lower levels of a system by the professional management also contributes to this process.

Above all, it is the active involvement of the Party and its cadres that is relied upon to give the necessary leadership to achieve unified planning. While the higher level Party organs lay down the broad principles and general lines of policy, the Party committees and cadres at the lower levels are not only expected to take interest in the working of water control systems, but to actively participate in their management to ensure that it has a unified perspective and that conflicts arising in the formulation or implementation of policy are effectively resolved.

The introduction of communal ownership and cultivation has no doubt meant a drastic reduction in the number of points at which conflicts could arise: Until 1979, the production team has been the basic unit of ownership and cultivation under the new system. It is several times larger than the average farm of pre-revolutionary times. Nevertheless, conflicts could and did arise between teams, and between brigades and communes, over the use of water and the allocation of responsibility for (and costs of) maintenance. In the past it was the large landowners and the gentry who either made and enforced decisions or had the power to influence the decisions of officials at the local level or even at that of the district and province. After the revolution, this role has devolved on the party cadres. Besides taking a lead in ideological education, the cadres are expected to participate actively at all levels of management. Political education and guidance of the professional managers is stressed; the committees of management at each level include functionaries of the Party at that level and, until recently, the key officials of the collective sector were also drawn from the political cadres.[32] On the one hand this mechanism is designed to promote a consensus regarding the overall objectives and guiding principles among the component units of the system; failing which it also serves as a mechanism through which the political sanction necessary to resolve conflicts and enforce decisions can be applied.

The Management of Maintenance

The Pre-Revolutionary Period

Historically, maintenance of large projects affecting extensive areas had necessarily to be undertaken by government. The officials responsible for the work, as a rule, organized and supervised the activities carried out with the help of corvée labour. By the nineteenth century however corvée had been formally abolished and the costs of maintaining public works had to be met from the general budget.[33] However, according to experts on nineteenth century China, the magistrates were not given sufficient funds to implement the task. They had to raise the required resources by means other than regular taxation. The customary way was either to recruit local residents to undertake the labour or secure contributions from them in proportion to the extent of their landed property. Sometimes a special fund was raised by means of donations when works on a large scale were required to be taken in hand (C'hu, 1962: 155).

Most studies on local government in the nineteenth century China suggest that the local officials had to depend a great deal on the gentry for the mobilization of resources as well as for the organization and supervision of the actual work. Since the labour service had been, by then, formally abolished and merged with the land tax,[34] this mobilization must have been an 'extra-legal' impost. If, as often happened, local officials were not diligent about their duties and/or the regional and provincial governments were indifferent, the villagers were largely left to fend for themselves.

Partly because of the lack of resources and partly the apathy of officials, a distinction had to be drawn between 'government dikes' and 'non-government dikes'. The following account of the situation in Kiangsi province by a local historian (cited by Hsiao, 1960: 286) is of interest in this context:

[Dikes were built] to prevent floods.... Some of them measured several hundred *chang*: others formed circles of one or two *li*.... After repeated breaks and crumbling, the usual practice was to petition the government for [repair] funds. Being afraid that the government treasury could not meet all the expenses, it was decided that all those that were listed in the old records be classified as 'dikes to be repaired by the government'. Whereas those constructed in later times should be classified as 'dikes to be repaired by the people.' Dikes in the latter category might be registered [with the magistrate], but the government did not inspect them... nor supply funds for their repair.... Hence appeared the classification into 'government dikes' and 'people's dikes'.

Non-governmental works constructed by local effort were maintained by local organizations. Again, we have few descriptions of the procedures in concrete situations and how they worked. In the case of the north China system referred to earlier the chief supervisor decided at the beginning of each year on the scheduling of repairs and indicated to his 'subordinates' the number of labourers required from their groups as well as the time at which and place where they are needed. The 'subordinate' was then responsible for mobilizing the labour from his group of farmers (each of whom was expected to contribute an able-bodied worker) and to ensure that the assigned work was completed. Disputes were generally resolved by the chief supervisor. The chief supervisor and his assistant were given a small amount of grain by every household on the basis of area irrigated (the rate being fixed by custom). The inspector was allotted a small plot of land by the villagers for his use. In addition, water fees were collected to cover other expenses for which a separate account was maintained by the chief supervisor. The levy was proportional to the area irrigated and the responsibility for collection vested with the subordinates.

In the 400 ha. system referred to by Hsiao the association of users undertook both routine and major repairs to the dyke and assessed contributions to be made by each farmer in proportion to the area of land irrigated. In Hainan (Motonosuke, 1979) all beneficiaries were expected to contribute labour for maintenance. The supervisors were responsible for mobilizing the workers. Water charges were collected by the management, the levy being borne equally by the landlord and the tenant. After deducting the necessary expenses, the balance was given to the supervisor of each sector. The state apparently did not pay any of the functionaries. There are also reported cases (Elvin, 1979: 91–2; Myers, 1975: 203–4) of local water management institutions owning their installations and 'sometimes lands or even mills to pay for their upkeep from rents and fees.'

We do have a detailed account of the sources and nature of conflict on the maintenance of a particular flood control system during the sixteenth and seventeenth centuries (Hamashima, 1980, 1981). Though it pertains to a rather distant past, the study provides such rare insights into the interaction between the state and local institutions in management of maintenance and the way in which this is affected by changes in land tenure, that its principal findings are worth recapitulation.

The terrain of the area studied (the Kiangnan delta in the Yangtze estuary) being very flat, efficient water control depended critically on maintaining the smooth flow of the rivers. This required periodic desilting of the rivers and creeks, as well as maintenance of the dykes in good repair. The dredging of the main rivers was a relatively infrequent operation (done once in '20–40 years') but the scale of work was large.

as the work was beyond the power of rural communities, whether individually or collectively, it was generally carried out on an extensive scale under the supervision of a high government official despatched specially for the purpose. Labour was supplied either... by supplying food to famine stricken people who in return were required to serve on water control works... or by the *li-chia* system... for recruiting corvée labourers. [Hamashima, 1980: 72.]

Under the *li-chia* system, resident landlords, who were generally appointed as local officials, mobilized the corvée labour and supervised the work of the peasantry.

Dredging the creaks and repairing the dykes, which had to be done far more frequently, often on an annual or seasonal basis, was managed locally. During the first half of the Ming period (fourteenth to fifteenth centuries) each landowner was responsible for repairing the full length of the dykes bordering his field irrespective of the extent of land held by him. The system appears to have worked fairly smoothly, in large part because the landlords owned most of the land along the dykes which, incidentally, was an advantageous location in terms of access to water and to the silt.

in the former half of the Ming, creeks and dikes in the Kiangnan delta were maintained on the initiative of the resident landlords who controlled water and related facilities. The *li-chia* system in which the tax chief and the *li*-headman were selected from among resident landlords was established on the premise that resident landlords were the ruling force in rural communities and thus functioned as an instrument for recruiting peasant labour for water control works. [Ibid.: 76.]

During the first half of the sixteenth century the dominance of the resident landlords began to decline as a result of the advancement of 'commodity economy' and a rise in land and corvée taxes. Their place was taken by landowners from the gentry who were mostly government offcials residing in urban areas. Not only were they not involved in farming but also enjoyed exemptions from corvée service. The resulting disintegration of the *li-chia* system and the vacuum in the leadership of water control activities led to a considerable deterioration in the facilities. In response to the resistence from the non-gentry landowners, attempts were made by the state to establish new regulations concerning water usage and to arrest the decline in cooperative and community relations.

First the partition of the *yu*'s was encouraged as this would bring a larger number of farmers within the ambit of the original principle governing the allocation of corvée. The second approach was to alter the basis of allocation of labour services (i.e. corvée) to a contribution based on the extent of area owned. This effort however came up against two difficulties. On one hand, since the gentry land was exempt and its importance was rising, the new arrangement meant an increase in the corvée burden on the commoner landowners. Even if the gentry's exemption from corvée were withdrawn or restricted this could not be enforced because they did not reside in the villages and did not have much labour power any way. In order to meet this difficulty it was proposed that the landlords would pay 'silver or rice as food costs for labourers and make their tenants provide the real labour' (Hamashima, 1981: 9).

In the event, attempts to modify the corvée system in the light of the new situation and bring about an improvement in water control maintenance met with considerable, often violent, resistence from the gentry. It took widespread peasant struggles (which took the form of tenants refusing to pay rent and rebelling against the gentry) before reformist elements in the bureaucracy and the higher levels of government began to actively promote the reforms referred to above during the late sixteenth and early seventeenth centuries.

Morita (1973) also cites a number of instances to show the progressive decline of the landlord's direct control over the management of water and the emergence of cultivators' organizations in its place. He suggests that the spread of absentee landownership and geographically scattered pattern of holdings led to a weakening of landlord involvement in water management during the seventeenth and eighteenth centuries, and to the emergence of new arrangements to take over the responsibilities (cited in Elvin, 1979: 95–6).

The Post-Revolution Period

For approximately a decade after the Revolution the fluid state of economic and political organization in rural areas together with the preoccupation with rapid expansion of water control projects resulted in a comparative neglect of maintenance. This, it became evident, was one of the factors reducing the effectiveness of the massive new investments during the fifties. Awareness of weaknesses in the continuing management and operation of water control projects, and the importance of ensuring that the facilities were inspected regularly and kept in good repair, is reflected in the general approach to problems of irrigation management and water use.

Construction is the basis; without it we might as well not talk about management. Management is the key. If it is not done well the project will not be fully effective; it is equivalent to reducing the amount of construction. Poor management may even lead to serious emergencies. [Nickum (ed.), 1961: 101–2.]

Project maintenance, involving 'inspection, maintenance and repairs at fixed intervals' is listed among the four major tasks of project management (the others being 'the rational operation of projects', the systematic 'observation of structures', and, interestingly, 'the improvement and rebuilding of projects'). At each level of irrigation organization, programmes for maintenance were expected to be drawn up in advance of each major irrigation season as an integral part of the programme of operations. These programmes were to be unified and coordinated by the body responsible for the project as a whole.[35] The responsibility for carrying out the work was also distributed between different layers. In general, the organization in overall charge of the project (such as the Irrigation District) was responsible for the work on the dam and the main canals. Work relating to the branch canals, distributaries, and local facilities (ponds, small drainage channels, field ditches, and pumps) was entrusted to organs of the collective sector, but subject to supervision and inspection by the officials of the district management.[36]

We have descriptions of the way maintenance was carried out in one or two projects. Thus in the Quianli irrigation project

The responsibility for annual repairs and regular maintenance on the main and branch canals was determined by management stations which delineated sections of land and percelled the responsibility for them out to various teams benefited according to the size of their irrigated areas.... The teams benefited by the small canals under the branch canals assume responsibility for their maintenance. [Nickum (ed.), 1981: 198–9.]

In the case of Shaoshan, which is a relatively large project, the irrigation district manages 'the maintenance of the main canals and their small-scale structures along *xian* (or municipal) boundaries. The *xian* (or municipality) subdivides its task by sectors among the local communes.' The collection and management of water fees is left to be handled by the *xian* or municipality. The district management 'merely coordinates with the communes in the district to establish good maintenance plans, formulate construction designs, carry out technical guidance, organize quality inspection and certification; properly handle the recording of labour time and settlements and patrol the canal for security ...' (ibid.: 183–4). This system, it is claimed, 'enhances the leadership of the local party committee over management work,' promotes activism among the masses in managing the canal, 'gives scope for greater flexibility and initiative, and gives the professional management departments', 'more time and energy to go out among the masses, investigates and studies, and carries on scientific experimentation, continuously raising the standard of management' (ibid.).

Labour, which is the most important resource and accounts for the bulk of the maintenance cost, is contributed by the beneficiaries. Besides supplying the labour needed for maintenance/repair of works within their jurisdiction, labour needed for works undertaken by higher levels were also to be provided by them. The general principle is that beneficiary units should contribute labour in proportion to area benefited. There are evidently cases where the allocation of responsibility between user units is based on other principles.[37] The water using units are expected to 'ensure that specific people are in charge of guiding the various items of work within the benefited area of the unit.' Production brigades and teams are required to 'divide the canal into sections fixing responsibility all the way down to specific individuals.' This system seems to work smoothly, though instances of conversion of open canals and drains into covered systems to save maintenance costs have been reported (ESCAP, 1978: 7).

Water charges are also levied on the production teams. The general principle appears to be that the charges must be adequate to cover the costs and if possible build a fund for expansion and improvement of the projects, but keeping in view the magnitude of benefits and the ability to pay.[38] The items of cost that are to be recovered through the water fees are explicitly defined. However, since the beneficiary units contribute labour for maintenance, it seems reasonable to suppose that water fees should cover the salaries of government officials assigned by the central management (not all of which of course is attributed to maintenance) and of material and equipment needed for repair and maintenance. The charges are on the basis of area, volume of water, or a combination of the two. The experience with volumetric charges, motivated by a desire to discourage excessive water use, has apparently been mixed: there is one reported instance of a project abandoning the experiment with volumetric charges after a while; two others have changed from an area based system to volumetric charges.[39]

The difficulty of collecting water fees fully was felt in the fifties and led to the formulation of certain general guidelines. The sensitivity of the matter is reflected in the implicit recognition of 'ability to pay' as one of the criteria in fixing the rates, the emphasis on development of sideline activities as a way of augmenting the incomes of the project management and hence 'reduce the water fee burden borne by the masses,'[40] and the fact that water rates decided by project authorities are subject to higher-level approval.[41]

Management of Water Allocation and Use

The other continuing task of irrigation institutions is to manage the allocation of water between different parts of the system and different users in particular parts of the system. This involves, (a) laying down criteria for deciding the timing, frequency and duration of water supplies to different sections; and (b) evolving mechanisms and procedures to regulate actual distribution in accordance with these criteria and to resolve conflicts that may arise in the process.

Historical Practices

Judging by the language of the early legal codes relating to the water management, it would appear that the allocation of water was subject to a high degree of government control. According to the code prepared during the Tang Dynasty (AD eighth century) 'all water uses were subject to government control'; no sluice could be constructed without the permission of local authorities; and for each canal and sluice-gate overseers had to be appointed. A common feature of these codes is that

they embody special rules and regulations concerning the distribution, utilization, control, and surveillance of water supplies made available by new water works. The timing of schedules of the waters in the new canals were fixed as well as the obligations of the users, the number of guards to be appointed to supervise and guard the water distribution... the obligation concerning the return of surplus flows. [Caponera, 1961: 259.]

on the local level, water distribution throughout the empire was under the strict supervision of permanently appointed overseers of canals and overseers of sluices in 'complete control of regulating the amount of water' at the time of flooding the fields. There groups organized at the sub-bureaucratic level were inspected by superior officials assigned annually by the prefectures. They had to show no partiality. [Ibid.: 256.] [42]

No private ownership rights in water were recognized, but only the right of use subject to specified rights of other beneficiaries. Even the 1912 Constitution, which was influenced by Western concepts of ownership, recognized only a qualified right of ownership over water. Caponera adds that

the main characteristics which have been found in the Han and the Tang periods remained practically unchanged up to the proclamation of the Republic in 1911. [Ibid.: 264.]

There are several reasons to doubt whether the government's role was in fact as wide-ranging in scope as the language of the formal codes suggest. The number of officials allotted to these tasks does not seem, at any time, to have been large enough to be able to attend to details of local level allocations. Even where officials were in position the local allocation problems appear to have been handled by local organizations without recourse to officialdom.[43] Moreover, as Caponera points out, 'in spite of written texts it was always recognized that custom should prevail when determining the extent of the characteristics of a water right' (Caponera, 1961: 273). [44]

The actual demarcation of water allocation functions between officialdom and local institutions as well as the nature of interaction between them is not known. It appears likely that the state officials were actively involved in regulating the use of water from the principal rivers and the distribution of water in large state-managed systems leaving the rest to be settled locally. State officials could not, or did not, pay much attention to projects constructed by landlords/clans for their own use and small community systems that served a few villages and were not part of a larger system. As to the manner in which the local organization managed water allocation, our knowledge is limited to a very few projects.

Elvin cites two instances where individuals owned water rights separately from their rights on land each of which could be traded independently of the other. One of the instances is a village in the nineteenth century where

watered vegetable gardens grew up with funds drawn from pawn broking and other trades in the city. Operations were handled by a garden guild. The ninety odd members held 'water shares'.... The channels and hydraulic installations were the common property of the guild.... Water rights might be bought, sold or borrowed separately from land. [Elvin, 1979: 89.]

The other instance is from north China in the sixteenth century where 'a spring that served 63 villages was monopolized by local landholders and those who needed water had to buy water tickets from them' (ibid.). How extensive such a combination of private ownership and commercialization of water was it is difficult to say.

The market system appears to offer greater scope in well irrigation: the area and the number of beneficiaries per source are generally smaller; moreover, there is much greater control over the timing and volume of water delivery. It is possible that the allocation by market played a more significant role in north China where groundwater irrigation is concentrated but there is no basis to judge just now important it was. In the case of surface irrigation works where the number of beneficiaries served per source is typically much larger and where the control over the timing and quantum of water delivery is much lower than with well irrigation, it is unlikely that the 'market' played a significant role as mechanism of allocation. Here some form of physical allocation, either under the control of those who 'owned' the water rights or through some form of community organization appears to have been the rule. There were evidently quite elaborate regulations and conventions concerning the timing and sequence of water use.[45] We have descriptions of two community-managed systems which, though far from adequate, give some idea of the water management problem in local systems and the way it was handled.

Fei (1939: 156–59) describes the management of a water control system in a village of the Yangtse delta some 130 km west of Shanghai. It has several distinct features characteristic of the region: the land is divided up by streams into *yu*'s varying in size (from about 1/2 ha. to 60 ha. in this particular village). Each *yu* is surrounded by water. Water for irrigation has to be lifted from streams along the banks of the *yu*'s and transported to its interior through ditches. Each pump irrigates a strip of land, dyked to hold water and laid out perpendicular to the bank. Since the fringes are typically at a higher elevation than the centre, the land within the *yu* has to be graded and dykes constructed parallel to the margin in order to avoid maldistribution of water between the fringes and the centre. (Without these dykes 'there will be a pool in the centre with the marginal land left dry'). In times of excess rainfall the *yu* had to be drained: a deep trench running through lowest part collects the excess water and takes it out of the *yu*. Relatively large *yu*'s have to be divided up into several segments for convenient drainage. These factors determine the basic layout of the irrigation and drainage channels as well as the cultivated plots. Individual farmers' plots are fitted into this broad layout.

In order to irrigate, water had to be lifted from the *yu*. This was done traditionally with the aid of a 'pump' made of wood and operated by human power. Each strip of land under a pump was divided into plots sloping to the centre and marked out by dykes. Each strip had a common ditch. Each plot in a strip had an outlet. At the time of irrigation, the ditch was first closed below its outlet and water allowed to run on to the plot. After irrigation, the outlet was closed and the ditch opened up to let the water in to the next field.

The first watering took place at the time of land preparation and again at periodic intervals, depending on rainfall, after transplantation and until the ripening of crops. Pumping for irrigation was invariably done on an individual basis with the aid of family labour but subject to several checks to regulate the distribution between plots and for sharing the labour involved.

People are not allowed to build dykes in the stream in order to monopolize the water supply. This is a common issue of dispute between villagers especially during drought. The water introduced on to the farm by human effort belongs exclusively to the person who has effected this by labour. The dykes are not allowed to be opened in order to 'steal' water from the higher plot. But a single plot may be owned by several persons. Each has a part in it. Since there is no dyke to separate the parts owned by different persons, the water is shared by all. In such a case, the labour spent in irrigation is equally distributed between the owners according to the size of the land in the plot. Most important of all, the level of the plot is maintained evenly in order that there should be a fair distribution of water. [Fei, 1939: 175–6.]

In contrast, the organization of work and division of responsibility for drainage was quite different, being much more collective in nature. In the process of drainage, the water was pumped out from the common trench of a *cien* (which is a segment of the *yu*). The people who work in the same *cien* share a common fate. They therefore develop a well organized system of collective drainage. Fei describes the organization for a 22 ha. segment of one *yu* with 15 pumps each requiring 3 workers.

The members of the *cien* are organized into fifteen teams corresponding to fifteen pumps. Each year one of a group will be charged with contributing the pump and managing the team. This position is taken in turn by the members of the group. Among the fifteen groups, there is a chief manager. This position is also taken in turn. At the beginning of the year, the chief manager calls the fourteen other managers to a meeting. A feast is prepared as a formal inauguration. The chief manger has authority to determine when the drainage should begin and stop. [Ibid.: 1723.]

Each farmer household with land in the unit was required to contribute labour for this operation in proportion to the area held by it. A system of standardized 'labour units' was also in vogue. The chief manager decided when drainage was needed and gave orders to the managers whose responsibility it was to mobilize the labour. Failure to show up for duty entailed specified material penalties (ibid.: 172–3).

The primary sources of conflict over water allocation in this system appear to have been attempts to break dykes in order to steal water from the higher plot and attempts on the part of one or more farmers in a particular plot to depress their portion to 'receive a favourable reserve of water' at the expense of others in the same plot. But Fei says nothing about the incidence of these disputes, or about how they were resolved.

In the case of the north Chinese village, it is reported that every spring irrigation water was let out after completing the necessary repairs to ensure smooth flow of water into the villages. Each village received its water 'at a prescribed time for daily intervals according to long standing custom [one whole day and night for every 250 *mou*]' on the completion of which the chief supervisor ordered closure of the sluice of that village and the opening of the sluice for the village next in line for water. When a village began securing water, the subordinates in that village informed farmers that they could irrigate. It would appear that the dredging of each section of the sluice was organized by the subordinate with the help of peasant labour as the waters begin to flow.

Again, nothing is said about the nature and incidence of disputes over allocation, but it is reported that 'when disputes erupted between households using an irrigation system' the chief supervisor 'immediately came to mediate and resolve the disagreement. Disputes rarely became so serious as to attract the attention of village headmen or local officials' (Myers, 1975: 201). In fact, 'the local administration maintained an office to mediate water disputes,' but at the time of the survey, in 1942, the office was practically defunct (ibid.: 209).

Changes Since 1949

The principal features of the present organizational structure for water management have already been described (see pp. 205–17) and attention drawn to several major changes from the earlier pattern. Here I am concerned specifically with the principles and procedures by which the allocation of water is determined and enforced.

Each irrigation district is required to prepare a comprehensive water use plan for its area every year and season.[46] The general procedure visualized is that before the start of the year (season) the production teams prepare advance estimates of their water requirements on the basis of the area that they propose to plant with different irrigated crops (which is itself related to the teams' production targets fixed outside the irrigation organization) and norms of water consumption per unit area (set by the central management on the basis of soil and climatic conditions and other relevant data, including the farmers' own experience). These are aggregated and considered by the next higher level of user units (brigades) before being passed on to the commune, and eventually to the central management.

On the basis of these submissions, the managers of the Irrigation district prepare an overall plan of operation, taking into account the requirements as estimated by the user units; the requirement for uses other than crop production (including water for desalinization and preventing salinization); effective rainfall; the likely availability of water from all storages and pumps irrespective of who controls them; and possible ways of correcting imbalances between supply and requirement. The committee or conference of representatives for the district as a whole has to approve this plan, but it is subject to review by the higher level political leadership. [47]

Thereafter, the water use plan is required to be translated into a detailed operational programme by quarters, or such periodic intervals as may be appropriate. This programme, in the case of the relatively large districts, consists of at least two levels: the central management decides the schedule of operation for the reservoir and the main canals, and also indicates 'the norms regarding the amount of water which can be drawn during the various stages' by the basic level water management units. In large systems this process may have to be gone through at several levels between the central management and the basic level units.

The basic level units then work out, in consultation with representatives of users, the details of the allocation plan between its constituent user units as well as measures for its execution. It also assigns the tasks among its personnel and those of its constituent units. The water using units are expected, with the basic level unit assisting in organization and training, to ensure that specific people are in charge of guiding the various items of work within the benefited area of the unit. It is expected that each team will have at least one full time person (at the rate of one for every 10–15 ha.) to attend to the task of maintenance and repair, and supervision of the actual watering of land; 'putting the irrigation plan into effect and rationally arrange labour and animal power, determining the order of irrigating shifts and systems for their relief and notify irrigation and patrol personnel of these;' 'promote rational irrigating methods and improve and elevate irrigation techniques;' and 'closely integrate irrigation with other technical measures for agricultural production' (Nickum (ed.), 1981: 251–3).

The procedure for preparing and implementing water use plans can be adapted according to local conditions and in the course of implementation. Management units at all levels are free to make marginal adjustments in the operational programme for their jurisdiction in the light of the evolving situation in the course of the season. Substantial changes in the programme (by way of changes in irrigation schedule, frequency, or amount of water or area to be irrigated) are permitted under certain conditions, e.g. breakdown in physical facilities, large deviation of actual from expected weather, and changes in irrigation sources. In such cases however the changes should be 'promptly' reported to both the higher level leadership organ and made known to the basic level management organization, water using units, and to all levels of the water allocation personnel. At every level the personnel in charge are expected to report to the higher level any changes in conditions warranting modification of the operational programme (ibid.: 256–7).

There are also general guidelines as to how water allocations are to be adjusted in the event that expected supplies fail to materialize:

1. Under normal conditions, water diversion and water allocation should be in accordance with an approved plan; this cannot be changed at will.

2. Irrigation districts with many kinds of water sources should unify the disposition of water sources, the allocation of water and the projects utilized.

3. When water sources are insufficient and rotational irrigation is carried out, priorities must be established: drinking water for people and animals; areas of severe drought; areas in which the economic benefits of irrigation are comparatively high, and areas that have been benefited for a long time.

4. For projects whose main purpose is irrigation, other uses of water should not affect the use of water for farm irrigation. During non-irrigation periods and when water sources are not abundant, water should not be released solely for the generation of electricity, processing, or fishing. (Ibid.: 255.)

The irrigation district is expected not merely to manage allocation but also (a) to promote greater efficiency of water use and reduce losses by introducing better techniques of irrigation and coordinated use of all available water sources in the area; and (b) to undertake programmes for new construction to augment the storage capacity in the command. These are in effect ways of coping with shortages of total water in relation to crop water requirements—the former by raising irrigation efficiency, the latter by increasing effective water supply. That significant improvement in either direction takes both time and conscious effort is evidently recognized.

The central management is expected to take the initiative in setting up research stations specifically to evolve better irrigation methods and techniques of irrigated farming adapted to local conditions, as well as to persuade the user units to adopt them. The management is also expected to play a leading role in constructing new storages and in modifying the original system (through realignment of the layout of the distribution network; lining the canals; improvement of drainage; consolidation and levelling of land) to improve its efficiency either by undertaking the works on its own or by encouraging and helping constituent units to take them up.

We have some idea of the kinds of problems faced in implementing this general approach to water allocation and use in particular situations and of the ways in which they were overcome. (See Nickum (ed.), 1981: Readings 3–7 and 10). The experience of one of these project (Meichuan) is summarized below.[48] Not only is it a remarkably successful effort in overcoming difficult problems, but the description of its experience is among the most detailed. Indeed, there are few accounts in the literature that discuss the technical and organizational aspects of irrigation management, the interrelations among them, and the way they evolved over time, with as much clarity and breadth of perspective.

The Meichuan reservoir located in Hubei province, completed in 1959, was designed to irrigate some 70,000 *mou* (approximately 4,600 ha.) out of a total cultivated area of 126,000 *mou* (85,000 ha.). The supply of water was much lower than planned, liable to large fluctuations, and altogether inadequate to meet the requirements of the command àrea. Actual irrigated area in 1959 and 1960 was around 50,000 *mou*. In the course of the subsequent 13 years the irrigated area increased to 120,000 *mou* and average yields per unit area doubled (Nickum (ed.), 1981: 82).

This was made possible by construction of additional storages, better use of storage and economies in the use of water. Between 1959 and 1972 new reservoirs (ranging between 100,000 m^3 and 10 m. m^3 capacity) with an aggregate capacity of some 23 m. m^3 were constructed. The restoration and improvement of some 6,000 small ponds in the command area added another 12 m. m^3. Taken together, this implies a 125 per cent increase in aggregate storage. Second, these storages and their distribution networks were linked to form the interconnected system to give greater flexibility in the allocation of water. In particular, it facilitated the use of ponds to meet much of the irrigation needs of early rice, leaving the main storage free for the second crop which needed more water and for which irrigation was more crucial. Third, waste in water use was reduced by rationalizing the layout of irrigation and drainage networks, by levelling and replotting the land in some areas, and by strengthening water management at the user level. Finally, the field water requirement was reduced by introducing improved techniques of irrigation, involving a lowering of the depth of standing water combined with periodic drying of the field.[49]

The implementation of all these involved sizeable new investment (all of which had to be mobilized from the users), far-reaching changes in the operating procedures, and a vast improvement in the organization and procedures for regulating water allocation. Besides technical problems, difficulties arose essentially from the fragmented perspectives of different groups (the professional management, the users, and the party) involved and conflicts among them. Thus, initially, the Irrigation District Management, manned wholly by professionals, tended to take a rather narrow view of its tasks and did not pay much heed to the local party leadership: it limited itself to the operation of the reservoir and the canals under its jurisdiction. The existence of numerous ponds in the command area and the possibilities of using the reservoir in conjunction with them was not taken into account. Nor did they concern themselves much with the way water was used.

Faced with the inadequacy of supplies from the original reservoir, they first explored ways of augmenting them and discovered in the process the potentialities of additional storages, including that of small farm ponds. To exploit these possibilities required fresh investment. The management persuaded the Party Committee of the District about the need for new construction, and the latter then organized the mass mobilization needed for the construction. The role of the professional management in the construction was a matter of some controversy owing to divergent views on the policy on water conservancy—a controversy that reflected political differences of a wider and more fundamental nature. It was resolved in terms of the general line laid down by the Party that the only way to achieve greater, faster, better and more economic results in farmland water conservancy was to go outside the management office to mobilize the masses in the commune sector to construct small-scale water works and persist in emphasizing the masses doing it themselves and integrating the large, the medium and the small. [Nickum (ed.), 1981: 87.]

It was further recognized that 'the Management Office had to manage properly not only the reservoir, but also the irrigation district's ponds' (ibid). In the process of implementing this line, the isolation of the professional management from the user was broken.

The management office organized survey and project initiation teams for small reservoirs and pond inspection groups to go out... to work with the cadres and masses in the commune sector, helping them train water conservancy technicians and whipping up a large mass movement to construct small farmland water conservancy works. [Ibid.: 88–9.]

As the small-scale works became more numerous and as experience in operating the system accmulated, the necessity for, and advantage of, 'unified management and unified control' became apparent. In the absence of an interconnected canal network, and partly due to the poor quality of works, it was found that the embankments in upper reaches tended to breach and too much water was used in the middle reaches while the lower reaches received very little. This led to the realization that interlinking the main reservoirs (to permit the use of surplus in some to cover deficits in others), linking the canals over the entire district, and linking the ponds to the canal (to expand the pond collection area and raise the 'restorage rate') was necessary. The experience also underlined the necessity of strengthening maintenance.

It however took some time, and the experience of managing a severe drought, for the Irrigation Management office to go beyond the problem of managing water release and consider those relating to the rational use of water. This led to a search for ways of more effective use of available storage to meet the water requirement of the two paddy crops over as wide an area as possible. Unified operation of the storage and ponds was found to require not only physical interconnection but also a clear set of principles as to when and how water from these different storages was to be released and a clear definition of responsibilities. Having established the operating principles (pond water to be used for early rice; and 'both reservoir and pond water to be used at the same time and closely coordinated' for the second crop with the ponds supplying most of the water in the later stage of the crop), it was found that its implementation would require a high level of management at the local level 'to do a good job of storing water in the pond' and of using them skilfully. This led to improvements in pond management by strengthening the system for fixing 'personnel, responsibilities, remuneration, and award and punishments.'

Conflicts, however, continued over how water was to be allocated rationally 'in order to guarantee the timely irrigation which promotes agricultural output and uses water sparingly....' Initially equitable distribution of water was sought to be achieved by allocating on the principle of 'near first and then distant'. When this resulted in excessive water in the upper reaches and acute shortage towards the tail end, the sequence was sought to be reversed, only to find that the upper and middle reaches could and, in times of scarcity did, interfere with the flow to secure their needs. Rather more elaborate procedures, aimed at relating water deliveries to the relative urgency of requirements determined on the basis of soil condition, the state of the crop, and actual soil moisture conditions in different parts of the command, have been introduced. (Nickum (ed.), 1981: 113–8.)

In times of drought even stricter procedures, involving joint inspection of actual field conditions and determination of quantum and timing of release, were invoked. The complementary changes in management organizations necessary to implement them (i.e. creation of mass water management contingents, training of peasant technicians, the creation of joint management committees, fixing of responsibility, etc.) have also been made. All these however did not eliminate conflicts among users over the sharing of water, especially in times of scarcity. This entailed additional measures: the upper reaches who tend to waste water should be required to use water sparingly; the middle reaches who had to block off water should be persuaded to take the entire situation into account and send 'unity water' (Nickum (ed.), 1981: 114).

The role of ideological education and mass propaganda in handling these and other allocation problems for the 'common good' is brought out by the following statement in the context of another project.

Repeated propaganda compaigns were conducted among the broad masses and cadres of the irrigation district... making a great effort to advocate the implementation of the spirit that only with high production does it (management) count... we clearly distinguish between having management of the irrigation district project rely on the masses and relying on a small minority; between basing policies on our own strength and solely state investment; and between being mindful of the overall situation and united in the management of water and profiting ourselves at the expense of others or just minding our own affairs. [Ibid.: 180–1.]

The official assessment of the experience of Meichuan and other projects place strong emphasis on the role of the Party and its cadre in creating and sustaining the necessary understanding among the broad masses and in providing the overall leadership to the water control management:

The key to how the Meichuan Reservoir could make large contribution though the reservoir is small, and do many things though the people are few, lies in Party leadership. [Nickum (ed.), 1981: 142.]

The Party's role is indeed pervasive: besides its role in educating the masses, its functionaries and cadres are closely and actively involved in decision-making and implementation at all levels of the irrigation organization.

In Meichuan, the Management Committee of the Irrigation district is headed by one of the secretaries of the *qu* party; a 'responsible commrade' of the Committee is 'assigned to oversee the work'. In the area of water management, the Party branch secretary in the management office is also a member of the *qu* party Committee which frequently reviews the work of the irrigation work. Similarly, the Party committee of the communes and the brigades are actively involved in the irrigation management committees at all levels: the Party secretaries are members of the management Committee, and at each level the irrigation organiztion reports periodically to the corresponding Party Committee. Personnel of the basic level management units are led by the commune party secretaries, while 'responsible comrades of the management sector and brigade party branch secretaries participate in the organization.' Those assigned for water management work at the brigade and team levels are also selected from cadres (Nickum (ed.), 1981: 144–5).[50]

By any standard, the management of water allocation and use in Meichuan is highly sophisticated and its reported achievements impressive. It however also appears exceptional. The very fact that Meichuan, and the two or three other projects whose experience have been documented, are held out as examples suggest that efforts to restructure irrigation institutions to achieve efficient use of irrigation supplies and their equitable distribution have not been widely successful. We have unfortunately no concrete information on the management of the 'representative' or 'typical' project in China. The following statement, though admittedly impressionistic, may more nearly correspond to the situation of the average project:

allusions are made in Chinese materials to a rift between supplier and user, characterized by a lack of ability on the part of the district to enforce the collection of water fees and by an absence of active involvement in each other's affairs.

One of the primary mechanisms to overcome the involvement gap is the 'irrigation district conference' held once or twice in a year, at which are discussed the repair and construction plans, water utilization plans and water-fee collection. That the communication is sometimes one way is indicted by a report in 1965 which claimed: 'In the past, when a meeting of representatives of the irrigated areas was called, either a demand was made for repairs to engineering projects or a collection of water charges asked for. The problem of production was basically ignored.

The most effective irrigated districts appear to be those which have direct ties to regular administrative apparatus, and where there is open exchange in terms of communication and involvement with beneficiary teams although water allocation plans and fee schedules are still set from above. [Nickum, 1976: 294-5.]

The last comment does suggest that the various ingredients of effective management emphasized in the general model advocated in the mid-sixties are indeed crucial. It also suggests that the ability to realize them in practice depends critically on a strong leadership whose function is not only to actively encourage the search for better solutions at the technical level, but persuade all concerned to accept their implication and also to intervene forcefully when necessary. In the context of post-revolutionary China, this leadership role, which is obviously in essence a political one, naturally fell on the Party and its cadres. It is however apparent that the Party cadres did not always play this role as effectively as in Meichuan. Also it appears likely that with the weakening of the principle of collective cultivation and sharing of produce that characterized the commune system, and the separation of the political from the economic functions of the commune since 1979, the party cadres' ability to play the kind of role they did in Meichuan, and is envisaged in the general model described earlier, will be significantly reduced. It is perhaps too early and, at any rate there is too little evidence, to judge how in fact the working of local irrigation organization has changed as a result of the 1979 'reform',[51] and what kind of new arrangements are emerging in its place.

SUMMARY

It may be useful, in conclusion, to recipitulate the principal features of the organization of water control in China that has been revealed in the foregoing review of the literature.

The Chinese water control system is both extensive and diverse in character. Currently, about 40 per cent of the total cultivated area is reported to be irrigated. For the most part, irrigation is from surface water drawn directly from the rivers or from small, local storages. Large-scale reservoir-based irrigation is relatively less important than say in India. Groundwater is tapped through shallow and tube-wells mostly in north China. Flood control works, the other important component of water control, consist of dykes, embankments, and retention lakes along major rivers and, increasingly, reservoirs. Most of the large reservoirs constructed in the course of the past three decades were primarily meant for flood control.

The character of the water control system shows marked regional variations: flood control works are relatively more conspicuous in the north China plains where they also tend to be on a massive scale. Irrigation is most widespread in south and south-east China, with practically all cultivated land being irrigated in some provinces. Small ponds, diversion works, and lift irriation from surface water sources are characteristic of this region. Irrigation in north China, the mid portion of the Yellow river basin (the 'loess' region), is largely canal-based, while in the lower reaches well irrigation is predominant. The proportion of irrigated area is also much smaller here than in the south and south-east. There is little irrigation in the west and north-west. Drainage, everywhere important, is particularly so in the lower reaches and estuaries of the Yellow, the Hwai, and the Yangtze rivers.

The weight of the available evidence suggests that during the two millenia or more of water control development prior to the Revolution, for the most part the state played a direct role in construction of relatively large-scale projects like the Yellow river flood control system. This remains true also of the post-Revolution period. Historically, these projects were constructed with the help of corvée labour. After the Revolution too a sizeable part of the cost of such large projects is being met by labour contributions generally from the beneficiaries of each project. However, in certain phases, especially during the Great Leap Forward phase, this principle was not observed and labour mobilization was more general. This, incidentally, is in striking contrast to India where there is no strong tradition of corvée labour for construction of water control works. Traditions of local resource mobilization for local works did exist in the past but have largely disappeared in recent times. Indeed, all large-scale (multi-community) systems, in the twentieth century at any rate, have been financed almost entirely by the state using hired labour. Even the local level investments (for field distribution, land improvement) to make use of water from large irrigation projects has mostly devolved on the state.

The bulk of China's water control (especially the irrigation) system, comprising small localized projects, has always been and continue to be planned and executed locally with local resources of labour and materials. The basis of mobilization and organization has however undergone radical changes. In contemporary China the scale of construction has been unprecedented, the functions of leadership have passed to the Party and its cadres; ideology plays a much greater role in mobilization; and the new system of land control and sharing of produce basically altered the relation between labour contribution and the benefits from water control investment. In principle, there should be great correspondence between the two now than in the past. Whether this has been so in actuality is somewhat difficult to gauge, partly for lack of relevant information and also because the principle of contribution according to benefits was not always observed. Furthermore, standards of design and construction suffered as a result of the scale of activity as well as the overemphasis on the 'mass character' of the programme thereby reducing the 'effectiveness' of the investment.

The organizational structure for maintenance and operation of water control systems, as in the case of construction, has been and remains considerably less centralized than is generally believed. Historically, while there was a sizeable hydraulic bureaucracy under state auspices, its focus was by necessity, and perhaps also by design, limited to the relatively large state projects. The peripheral parts of these large systems as well as the smaller, essentially local, systems were managed locally. The institutional arrangements varied a great deal, ranging from individual ownership and control to community management. There is reason to believe that, directly or indirectly, the large landowners, clan heads, and gentry bureaucracy, who were the centres of economic and political power in rural areas, were the key figures in the management as in the construction of water control projects. These traditional organizations had fallen into disuse for several decades preceeding the Revolution.

After 1949, private rights in land and water have been abolished and vested with the state or the collective sector. The organization for maintenance and operation has been streamlined. The overall responsibility for the management of a system now vests with the 'lowest administrative level whose boundaries encompass the benefited area.' In larger multi-commune projects professional managers, often appointed by the state, play an important role. The use of specialized personnel trained in techniques of water management is stressed at lower levels and in smaller systems. The typical project is managed by a multi-tired organization, but at all levels the active participation of users and managers under the leadership of the Party is stressed. This again contrasts with the situation in India, where partly because of predominance of large surface projects, the state bureaucracy plays a much more extensive and pervasive role in management. The extent of involvement of users in managing even the peripheral components of large systems, not to speak of the sytem as a whole, is very limited in comparison to that China.

Prior to the nineteenth century, maintenance of projects under state management was done with the help of a corvée system of labour mobilization. Even after it was formally abolished (in the nineteenth century), the practice appears to have continued because the funds allocated for the task were generally inadequate. Small projects and local components of large systems were maintained locally with labour contributed by beneficiaries. We, however, know little about how well it worked. Water control maintenance was intimately linked to land tenure and the local power structure, both of which had a bearing on the principles for allocation of costs among the beneficiaries and the way in which the principles were enforced.

In the post-revolution period too beneficiaries are generally expected to contribute labour for maintenance in proportion to benefits. The task is divided between different levels of the management. There are regulations that require each system to prepare a comprehensive maintenance programme and to allocate responsibility for its implementation to designated personnel at each level. Water fees are levied to cover costs of maintenance other than labour, the cost of management, and sometimes to build a fund. In some cases they are also sought to be used as a way of enforcing economy in water use. But as elsewhere, water rates are a sensitive issue: attempts to raise them to meet rising costs face resistence from users, and such revisions are apparently subject to higher level approval.

Traditionally the Chinese law, which was codified as early as AD eighth century did not recognize private ownership rights over water. People had only *rights of use* over water subject, however, to specified rights of other users from the same source and subject to customary practice. It was only in the 1912 Constitution that ownership rights were formally recognized, but even this was qualified. Instances of purchase and sale of water rights for a piece of land separately from the right to land itself have been reported, but this seem unlikely to have been widespread. The more general practice appears to have been some form of physical allocation based on custom and tradition. In some state works, the allocation of water from a particular diversion and sluice-gate was explicitly incorporated in formal codes. Judging by the few recorded instances, there were elaborate, if unwritten, conventions concerning the rights and obligations of users, the basis on which the sequence and duration of supplies to different users was to be determined, and the procedure for resolving disputes.

The role of irrigation management has been greatly broadened over the past three decades. Its functions are no longer viewed as merely those of formulating and implementing schedules for release of water to different parts of the sytem. Important as these functions are, the irrigation organization is now also expected to play an active role in identifying ways of augmenting water supply and making more economical use of it, persuading users to adopt them and undertake to implement the improvements. The experience of a few specific projects shows the difficulties that arise in giving effect to this conception. Conflicts arise between professional managers and users, as well as among users. Resolving these conflicts and evolving rational and yet generally acceptable solutions is evidently a difficult process whose success requires the Party's active and sustained leadership. Remarkable success is reported in a few projects whose experience has been documented.

POSTSCRIPT

At the time I wrote this chapter Chinese agrarian organization was in the process of being radically restructured. The abolition of communes, introduction of contract farming on an individual basis, and the diminished role of party cadres were expected to affect the functioning of water control institutions. It was however too early to assess the nature and extent of this impact. Unfortunately even now, nearly two decades after the reforms, we do not have a comprehensive picture of this aspect. It is however possible to piece together a broad picture from papers presented at a recent international conference on Irrigation Management Transfer (Vermillion et al. (eds), 1995). At this conference, held in Wuhan in 1994, a number of papers dealt with the post-reform developments, the reforms being attempted, and the experience of specific systems in different parts of China. Additional information could also be gleaned in the course of discussions of these papers. The following account is based on this material.[52]

The total irrigated area is now placed at 53 m. ha. of which 50 m. ha. are said to be 'effectively irrigated' by some 86,000 storages of varying sizes (with a total storage capacity of 468 bcm), 3.35 m. tube-wells, and

0.49 m. pumping stations. Nearly a fourth of the irrigated area is estimated to be served by wells and pumping stations. The figure of total irrigated area is substantially higher than the estimate for 1975:43 m. ha. (Vermeer, 1977).

Small reservoirs, ponds, and pump stations irrigating less than 667 ha. (or 10000 *mou*), and managed by village communities, account for 27 per cent of the irrigated area. Government managed irrigation districts (mostly those commanding over 667 ha.) account for 47 per cent of irrigated area. Of this, large systems (commanding over 20,000 ha.) account for only 8 m. ha., i.e. about a sixth of the total.

The functioning of irrigation systems was disrupted in the wake of the agrarian reforms. Even as individual cultivation increased, and with it the demand for better quality irrigation services, farmers were reluctunt to participate in maintenance of facilities. Amidst the confusion about the roles and responsibilities of individuals, villages, and officials under the new regime, it was not possible to get farmers to contribute to maintenance or to check tendencies to steal water and damage facilities. Conflicts increased and facilitates deteriorated. The irrigated area declined.

All this prompted a major reorganization of water control institutions. There appears to have been no major change in the categorization of systems (central government, provincial government, and locally managed), or in the division of functions and responsibilities between different layers of management. However the jurisdictions of irrigation districts have been more clearly demarcated; they have been legally empowered to take action to protect the facilities against encroachment and damage, and ensure proper use of water and soil resources. A systematic attempt at verification and accounting of assets, and identification of ownership, has also been initiated.

A detailed survey of large systems was undertaken to assess the present condition of the facilities and the rehabilitation works required. The survey showed that reservoirs, canals, and structures had suffered significant deterioration and that they were operating well below their capacity.

The managerial organs at all levels have been thoroughly recast. Earlier, government officials and party cadres were prominent in the management committees at all levels. At the lowest level the new system envisages replacement of the production teams with village irrigation management groups consisting of farmer representatives elected by the irrigators. These groups are required to prepare and implement the annual plan for maintenance, manage irrigation, and collect water fees. The induction of elected representatives of users to management committees at all levels of irrigation systems (including the top level policy-making boards) is the second principal feature of organizational reform. There is strong emphasis on participatory, co-management of irrigation.

Contract management is a major innovation, whereby the responsibility for carrying out various tasks related to irrigation is delegated to individuals and organizations. The contractors are apparently selected through competitive public bidding and appointed for a few years at a time. The contracts specify the tasks to be performed, and the norms of expected performance with regard to specific indicators. Competition in bidding and the fact that contractors who do better than the agreed norms are paid extra (or are allowed to keep the gains) even as underperformance attracts penalty, are expected to reduce costs and improve efficiency.

The papers also refer to a plan for assessing the value of facilities in each irrigation system and assigning property rights to the state, irrigation organization, and irrigators on the basis of their respective contributions to building the facilities. The idea appears to be one of converting the systems into something analogous to corporations or cooperatives vesting nominal ownership of the shares with respective entities, along with the right to a share in the profits. This is expected to induce irrigation organizations to mobilize more of their own resources for rehabilitation and improvement, improve management methods, and ensure more efficient use of water.

A reform of the system for water charges seeks to rationalize their basis on the principle of recovering operational and maintenance costs as well as contributing to the costs of rehabilitation and improvement. The new system aims at replacing the earlier area-based system by a two part tariff consisting of a flat charge related to area and a variable fee related to the volume of water used. Along with this, the irrigation organizations are encouraged to develop so-called 'sideline enterprises' to generate resources to meet part of their costs, presumably because this reduces the extent of increase in water charges required to break even.

Another innovation is to lease management of facilties to a third party for five or more years giving leaseholders the permission to set up sideline enterprises and use a part of the earnings for irirgation development in the hope of avoiding a hike in water charges. There is however little information on the scope and terms of the lease, the relation of leaseholders vis-à-vis users and employees.

Several of the problems of Chinese irrigation faced in the post-reform phase (poor management, excessive reliance on bureaucracy, overstaffing, inadequate cost recovery, and resistance to increase water charges) are not new. As will be evident from the earlier part of this chapter, most of them were reported in the pre-reform era. Some of the solutions, e.g. system rehabilitation and improvement, better water management, and sideline enterprises, have been attempted earlier. I have also referred to detailed accounts of some apparently successful experiments along these lines.

The major difference of the system is in the organizational and managerial changes being currently attempted. However, the scale of these changes and their impact cannot be judged from available information. The specific examples, cited in the Wuhan papers cannot be taken as a representative sample. In any event, even as they report signficant improvements in the wake of reform, they also cite the difficulties encountered.

Thus, there are references to the limitations of contract management in the context of efforts to define property rights; contracts being 'subjective' and 'unrelated to economic law', and the lack of clear relation between contract targets and improvement in management. Solutions to the problems of dealing with officials already on the rolls of the irrigation organizations remain elusive. Resistence to raising water rates continues to be indespread. Problems of collecting water dues appears to persist. The papers and presentations by Chinese experts tended to stress the experimental nature of the reforms, their willingness to experiment with different alternatives and their awareness of the problems and the need to learn by trial, error, and experience. The future course of the Chinese attempts at reform, their implementation and impact on the performance of irrigation systems, and the quality of water management, merit close watching.

ACKNOWLEDGEMENT

This chapter in its original form was written in 1983 when I was a Visiting Research Fellow at the Institute of Developing Economies, Tokyo. It was published under the VRF Monograph series of the Institute in 1984.

NOTES

1. Japanese scholars have done much detailed work in this field but, unfortunately, very little of it is available in English. Some idea of its scope and content (and especially of Akhira Morita's work) can be had from Elvin (1979).

2. Since no translation of Morita's book was available, my references are based on Elvin's review of it (Elvin, op. cit.). This is hardly satisfactory but I cannot do any better under the circumstances.

3. Some of Hamashima's work is published in English: Apart from a paper in

the *Acta Asiatica* (1981), an English summary of his 1980 book is also available.

4. At the time of revision (April 1983) I came across Nickum's paper on Chinese Water management prepared for the World Bank. This paper is almost exclusively concerned with contemporary China, while mine covers the pre-revolutionary and post-revolutionary periods. The prespective of this review is also somewhat different: the intention here is to bring out the contrasts between pre- and post-revolutionary China, and in the context of an attempt at a wider comparative study of irrigation institutions in relation to agricultural production.

5. According to Kojima (1982), the aggregate storage capacity of all reservoirs other than small ponds is 400 bn. m^3. Assuming an average of capacity of 50 m. m^3 for medium reservoirs (10 to 100 m. m^3) and 2 m. m^3 for small storage (1 to 10 m. m^3) I estimate the storage capacity of large reservoirs at 140 bn. or 35 per cent of the total. The capacity of the 6 m. ponds (100,000 m^3) could be 60 bn. m^3. According to an earlier study (Kojima, 1975), there were only 128 projects irrigating over 20,000 ha.

6. King (1911: 297–301) refers to some six types of traditional water lifting appliances used in China. These include the 'swinging basket' 'portable spool windlass', 'quadrangular cone shaped bucket and sweep', 'foot operated water wheels', animal drawn power wheels, and wheels driven by the current of the river. (See also Needham, 1971).

7. In 1965 'power driven irrigation and drainage systems' served 84 m. *mou*, or a little over 16 per cent of the 520 m. *mou* irrigated and drained by all means (Kuo, 1981).

8. This section draws heavily on Traeger (1980) and to some extent King (1911).

9. The Yellow river has one of the highest silt concentrations in the world: its waters carry about 34 kg silt per m^3 in comparison to 10 kg by the Colorado and 1 kg by the Nile. The total volume of silt carried by this river is placed at 415 bn. m^3 per annum (Traeger, 1980: 220).

10. For details see Chao (1970: 135–6) and Vermeer (1977: Ch. VI).

11. These data are from Traeger. Other sources (Chao, 1970; Vermeer, 1977) give much lower estimates of double-cropped area (under 40 per cent).

12. In this context the following statement by Chao (1970: 124) is of interest.

Traditionally Chinese peasants and the local governments were so keenly concerned about the scarcity of agricultural land that irrigation systems normally had been built with a minimum use of land. In south China farm ponds were common.... Some ponds were isolated whereas others were connected by a few short leading channels. Farmers used water wheels and other instruments to carry water from ponds or channels into the field. Cases of exclusive reliance on the force of gravity to distribute water were not common because this type of irrigation used too much land. In the North subterranean irrigation with either shallow or artesian wells was important. There was a strong aversion to reservoirs.

13. This and subsequent references to Motonosuke are based on a rough summary by courtesy of Dr. Kojima of the salient findings reported in Motonosuke (1979: 466–741).

14. Cited in Elvin (1979: 95–6).

15. According to Hsiao (1960, 183):

> The Imperial government did not undertake to construct irrigation slucies or dykes in the villages but encouraged such work by giving legal protection to 'water benefits' resulting from private efforts.

He also cites a number of examples of such works undertaken at the initiative of local inhabitants and argues that

> It is quite certain that irrigation work, especially that which involved organized effort was promoted by landowners. It is highly probable that some of the promoters belonged to local gentry families or powerful clans. [Ibid.: 284.]

Chang (1955: 59) cites an instance where

> a gentry member of Chekiang was so interested in irrigation problems in his own and neighbouring areas that he travelled extensively through many localities of Kiangsu and Chekiang and carefully worked out the causes of flood and drought. He worked out detailed plans for correcting the situation throughout the entire area pointing out the places at which rivers should be dredged and at which places water should be conserved.... His plans were taken seriously and were used on several occasions when actual construction was undertaken.

He also gives details of some 12 projects in one prefecture from which it is clear that most of them were the results of petitions made by the gentry to the Hsien magistrate and in some cases to the provincial governor.

16. Most of the subsequent material is pieced together from Donnithorne (1969), Nishimura (1971), Vermeer (1977), Kojima (1977), and personal discussion), Greer (1979), and Nickum (1979). Of these Vermeer's is by far the most detailed and informative on the construction of water conservancy works and in particular of the techniques of mass mobilization and their consequence. All the above sources attempt to reconstruct the Chinese experience on the basis of a painstaking collation of news, articles, and editorials from Chinese newspapers and periodicals dealing with particular projects, or aspects of them, in different regions. (In this context, also see Carin, 1963.) There appear to be very few primary Chinese sources dealing with questions relating to water conservancy from an overall perspective. The two significant exceptions are a 1958 publication in English entitled 'China's Big Leap in Water Conservancy' and an English translation (Nickum (ed.),1981) of Chinese accounts of some 5–6 specific projects. I have cited these two sources wherever relevant.

17. According to a recent Chinese estimate, the total investment in basic irrigation and water conservancy works between 1949 and 1978 was of the order of 130 bn. *yuan* of which 76 bn. represents state expenditure and the balance

contributed by people's communes and farmers (cited by Kojima, 1982: 401). He estimates that people's contribution (mostly labour) amounts to an average of 12–13 days of 'Semi Compulsory labour' per worker per year during this period as a whole. Including other kinds of work, the average is likely to be around 20 days per worker per year.

Another recent report (ESCAP, 1978: 6) suggests that the state gives subsidies, mostly in the form of scarce materials, or materials not avilable locally, amounting to 10–20 per cent of project cost for 'better-off' communes, and 30–40 per cent for the poorer ones.

18. The general approach is brought out by the following statement in the 1965 official publication *Irrigation Management*

> The purpose of building the water conservancy works had been clearly explained to the peasants at the very beginning. They all knew that the work was for their own good and for the good of posterity. They knew, therefore, that it was not right to ask for appropriations from the government and that they should count on their own efforts in building the project. Subsidies were, however, obtained with regard to a couple of medium reservoirs. As to all other works, they were built at the expense of the coops. While at work they brought their own food and used their own tools. Wherever there was a shortage of material they would increase their investment in the coop which would then buy for them. [Nickum (ed.), 1981: 4–5.]

For an excellent discussion of the scale and system of labour mobilization in different periods, and the problems involved, see Vermeer (1977: Ch. II). The issue is also discussed in Nishimura (1971), and Nickum (1978).

19. These estimates ostensibly relate to work done by peasants on projects inside and outside their community (Vermeer, 1977: 61) while the estimates cited for 1950–7 are said to concern only state-led projects. This would imply that the decline was greater than suggested. Nickum's (1978: 280) estimates for 1964–5 and 1965–6 are about three times those of Vermeer but suggest much the same direction of change.

20. For a discussion of the concept of 'Irrigated area' underlying Chinese statistics and the problem in interpreting statistics see Dawson (1966: 152–66); Chao (1970: 121–2); Vermeer (1977: 183–7).

The following points are noteworthy:

(a) Prior to 1956 there was apparently no standard definition of irrigated area.
(b) According to the definition adopted in 1956, 'irrigated land *refers to* cultivated land that is provided with water through fixed and permanent irrigation facilities such as channels, reservoirs, ponds and dams, wells; water wheels and pumps' (Vermeer, op. cit.: 183) Wet fields that did not have 'fixed and permanent' facilities were supposed to be excluded from the category of 'irrigated land'.
(c) During the Great Leap Forward phase 'irrigated area' also referred to areas that were within the 'control range of trunk canals, wells or irrigation

pumps for which, owing to lack of feeder lines, land levelling or lift irrigation no irrigation was yet possible' (ibid.). The concept would correspond more to 'irrigable area' rather than area actually irrigated. This, together with possible over-reporting, may account in part for the unusually large rise in irrigated area during the late fifties even as the subsequent decline may be due in part to stricter definition.

(d) The 1956 redefinition also sought to introduce greater precision in the concept of 'irrigated area' by specifying certain standards of water supply (in terms of the capacity to withstand droughts of specified duration that varied from region to region). It is not however clear how far this is reflected in the statistics. The category of 'stable high yield land' possibly reflects this concept, but the data are not reported on a regular basis.

21. An official publication from Yunan province (People's Pub. Co. of Yunan Province (ed.),1981) is of interest in this connection. The contents of it were summarized for my benefit by Dr. Kojima.

22. Kojima (1982: 402) writes:

One of the biggest changes in the new agricultural policy is the decision to do away with such collective labour. With the increasing prevalence of the system of production by farmers on a contract basis, the right to make decisions regarding the use of labour has been transferred from the people's communes to the production teams and individual farmers. Such labour mobilization can only be accomplished now by economic means (with compensation), it will cost a good deal in the future.

23. Myers' (1975) paper is based on material collected by Japanese researchers during the early 1930s. A six volume report (in Japanese) based on this research was published in 1958 under a title that translates as 'A Survey of Traditional Rural Customs in Chinese Villages'.

24. See Vermeer (1977: 279n) for the principal provisions of the 1950 nationalization measure for one region.

(In Meichuan (Nickum (ed.), 1981: Reading 3):

The main reservoir and five smaller reservoirs are managed by specialist organs; sixteen very small reservoirs are managed separately by specialists sent from beneficiary communes and brigades; and ponds are managed by production teams.

25. In the Shaoshan project (ibid.: Reading 5) the responsibility for maintenance and repair is divided by county/commune boundaries and not by project components.

In the case of the Yingjing canal, there are 4 levels:, the irrigation district management bureau responsible for the management stations whose jurisdictions are defined by the 'layout of the canals and distribution gate systems' (not necessarily coterminous with main or branch canals). Below this level, the division is on territorial lines (Ibid.: Reading 7).

In the Tsing Ping project (see ESCAP, 1978), the reservoir management division is responsible for maintenance and operation of the reservoir, distribution, and drainage systems down to the level at which the responsibility is assigned to the communes. Distributory canal committees are organized on a commune basis.

26. We have information on the constitution of the Committee only for one brigade level project (Tong Hsin Brigade). It is managed by a 9 member team of whom 4 (including the secretary, the brigade leader, and accountant) are nominated by the commune and the rest are elected. (ESCAP, 1978: 40).

27. Insofar as water management officials of the collectives are selected from leading party officials, and this appears to be the general practice, the first and the third categories get merged. Thus, in the Tsing Ping Project (serving 11,000 ha.)the Reservoir Management Committee consists of the Vice Chairmen of the 7 Communes/townships in the project command area (who were senior members of the commune Party Committee) and the Deputy Director of the Reservoir Administrative Division. The latter, a professional appointed by Government, is Chairman of the Committee.

At the next level are the sub-main canal management committees which consist of all the deputy brigade leaders (who again are generally key party officials) served by the water management stations. The officials of the Reservoir Administration Division in charge of each station is also a member of the sub-main canal Committee.

At the brigade and the team levels all water management personnel (including brigade level committees) are drawn from among their members (ESCAP, 1978: 30).

28. It is perhaps useful to note that there are two broad categories of personnel involved in water management.

(i) The personnel who comprise the management committees at various levels.
(ii) The personnel who are responsible for carrying out the decisions of these committees.

The latter category in turn includes, (a) The technical and other staff employed by the central project management; (b) the personnel of the collective sector assigned for water management work at the team, brigade, and commune levels; and (c) personnel (of whom pump-set operators and mechanics are instances) employed by the commune and brigades to carry out water management tasks requiring particular skills.

It is difficult to get a comprehensive picture of the total number of personnel in each of these categories for any of the specific projects described in ESCAP (1978) and (Nickum (ed.), 1981). Nor do the norms suggested in the 1965 manual on Irrigation Management (Nickum (ed.), 1981: 66) give an accurate indication because they relate to technical/administrative personnel to be employed it the system level; they also do not cover all categories of technical personnel. This appears to be *the reason* why the norm for large projects (1–3 persons per 10,000 *mou* of systems irrigating over 1 m. *mou*) is much less than for small system (3–8 personnel for a project serving 10,000 to 100,000 *mou*). In the former case, additional personnel will be required to undertake the technical/administrative tasks at the commune level and below. Also most personnel in the collective sector appear to be drawn from among the members and not employed from outside. The team level water management personnel are clearly not included in the norms. The general standard seems to be one person per 100–50 *mou*. Incomplete information for one commune level project (ESCAP, 1978: 40) shows that, including team level personnel, about 116 persons were involved in management for an area of some 2,000 ha. which works out to an average of 6 per 100 ha.

29. The 1965 guidelines had this to say on the subject:

> The porfessional irrigation crews... are the basic level water utilization organizations of the mass management organization. They play a very important role in management because they are the direct implimentators of the water utilization plans directly affecting the quality of watering. [Nickum (ed.), 1981: 70.]

The collective sector is said to have two categories of personnel: the year round irrigation crews and seasonal crews. The former attend to water management during the irrigation season and to maintenance during the off season. Based in production brigades, the personnel (at the rate of one per 100–250 *mou*) are drawn from the teams in accordance with area benefited. Seasonal crews, operating under brigade or team leadership, work on irrigation during the watering season and on agriculture the rest of the time.

30. This criterion is referred to in a number of projects. For instance in Shaoshan:

> Each commune... has provided one or two cadres to take charge of managing the work of the canal system. The vast majority of these comrades are selected from among the important responsible cadres of the production brigades; their political qualifications are good, and they have practical work experience, can relate closely with the masses and constitute the backbone for the management of the irrigation district. [Nickum (ed.), 1981: 191.]

31. The way this is suposed to work is described in 'Provisional Measures for the Planned Use of Water' (Reading No.11 in Nickum (ed.), 1981). See paras 108–111.

32. The importance of party leadership is stated in general terms in the introduction to 'Irrigation Management'and in the 'Provisional Measures for the Planned Use of Water' (Nickum (ed.), 1981: 61; 252), but the various features mentioned here come out explicitly in the description of individual projects.

33. The availability of funds was however not the only problem. The efficiency of maintenance of state-managed projects also depended on the general level of the administration. Thus Perkins (1966: 172), referring to the independent authority set up during the Ching dynasty to manage the dykes along the Yellow river, notes:

> when this authority was at its peak officiency, it together with vigorous local officials, were able to prevent major flooding. But when corruption became rampant in that administration, the dykes were neglected and the silt laden river broke out and inundated large areas of farm land.

34. The labour service tax was in theory imposed on male adults aged between sixteen and sixty in *lieu of labour service*. This was introduced in some parts of China from the early eighteenth century but apparently did not become a general practice till the mid-nineenth century. See Ch'u (1962: Ch. VIII) for a discussion of the taxation system and the way it worked.

35. The management of each irrigation district is expected to 'unify the formulation of programmes for repair, improvement and expansion of projects and is expected to carry out regular maintenance and renovation and improvement at fixed intevals' (Nickum (ed.), 1981: 67–8).

36. The basic level management organization (i.e. 'the main, branch and sub-branch canal main management committee or the irrigation management organization of the peoples communes or State farm') is expected to help the production brigade and production team establish sound watering and canal maintenance organization...' and 'organize the water using units to repair the canal structures properly...' (Ibid.: 253.)

37. In the Shaoshan project (16,000 ha.) the maintenance and repair responsibilities are subdivided by county/commune boundaries and not by project components, leading to unequal sharing of the costs (including labour contributions) among teams. (Ibid.: 182–3). This appears quite exceptional.

38. A 1958 report (See Nickum (ed.), 1981: Reading 12) states:

> the standards of water fee collection are determined on the basis of expenditure for project upkeep and maintenance and management cost and also in keeping with production increases and the capacity of masses bearing the burden. Funds may be accumulated in suitable amounts and used for improvement and expansion of projects. The amounts to be collected may be worked out by the provinces, the autonomous regions and directly administered municipalities concerned.

Also significant is the provision that the water fee 'should not be treated as local financial income; nor should they be transfered for use as expenditure on non-water activities.'

39. In the case of Meichuan, it was found that charging by volume led to a situation where some believed that they should get as much water as they asked for irrespective of others needs and the overall situation; upstream users instead of requesting supplies from the canal began to use seepage; others were reluctant to ask for canal water, even when they needed it. Because of this the system was abandoned in favour of a system of area-based charge combined with measures to propagate economical methods of using water and enforcing equitable distribution. (Nickum (ed.), 1981: 148–9.)

In the Quianli canal on the other hand, the practice of charging on area basis 'regardless of whether they used water or how much water they use 'was found inequitable. So the system was changed whereby' in addition to a small basic water fee based on irrigated area, all those who benefited pay their main fee according to actual amount of water they use.' (Ibid.: 198.)

It is possible, as Nickum suggests, that this difference reflects the fact the Quianli had less direct control over water supply than Meichuan.

40. This is also stressed in subsequent (1972) regulations concerning a particular project (see Nickum (ed.), 1981: Reading 13). Further, Nickum (1976: 293) reports that in some cases increase in water rates have been avoided by reducing the paid professional managerial staff and replacing them with personnel of the collective sector.

41. The 1972 regulation referred to above specifically lays down:

The mass water management organizations should discuss and draw up standards and methods for collecting water fees. *This should also be submitted to the responsible departments for approval and put into effect.* [Emphasis added. Nickum (ed.), 1981: 267.]

Nickum also reports that in the case of the Red Flag Canal, the annual decision on water rates by the project organizations is subject to *Xian* government's approval.

42. The following passage (Caponera, 1961: 255–6) is also relevant:

Whenever fields are to be watered, prior notification of the area should be given, and they [the people] should take their needs in turn. When the water has reached everywhere, the flow should be blocked. It is the first duty of the authorities to ensure that water has equitably spread everywhere....

43. In both the Hainan systems referred to by Motonosuke (1979) and the north China system described by Myers (1975), there are references to the existence of local officials for mediating disputes. Myers (1975: 201) says 'Disputes rarely became so serious as to attract the attention of the village headman or local official.'

44. On the concept of water rights in Chinese law see Caponera (1961: 254–5, 268–70).

45. The following statement based on the Tang dynasty code is of interest in this context:

The time distribution schedules were calculated and set by users themselves in minor irrigation districts under the supervision and inspection of the water officials; in the larger ones, the law provided for the period of water use and distribution for irrigating the fields....

46. This description is based on the 'Provisional Measures for the Planned Use of Water' issued in 1965 by the Ministry of Agriculture 'to provinces, autonomous regions and municipalities for trial implementation' (Nickum (ed.), 1981: 245–58).

47. 'After the conference of the representatives of the Irrigation districts or the irrigation management committee has discussed and approved the plan, it should be reported to the higher level leadership for review and approval' (Nickum (ed.), 1981: 247).

48. The Meichuan experience is the subject of Reading No. 3 in Nickum

(ed.), 1981. By and large, the experiences of the other projects reported in this volume are similar, especially with regard to problems and conflicts over reform and the importance of the Party leadership.

49. Unfortunately the document does not give any idea of the actual irrigation water consumption (aggregate or per unit area) in the later years. If there were no change in this coefficient, total irrigation water use must have (on the average) risen in proportion to irrigated area which has more than doubled from 1959–60 to 1972–3.

50. The role of the party in successful management is explained thus in the context of another project (Shaoshan):

> Water management must be placed under the unified leadership of the party. The Provincial and prefectural committee pay great attention to the project and regularly direct our work. The main leadership comrades of the prefectural and provincial water conservancy and electric power bureaus also assume leadership responsibilities in the Irrigation district management. The various levels of party committee of the irrigation district have also incorporated in their agendas the management of the Shaoshan Irrigation Bureau. In addition there are important responsible comrades who have divided the management duties of this aspect of the work, thus strengthening the actual direction. They arrange, inspect and sum up and exercise leadership with regard to key problems. We especially educate the people... helping them appreciate the spirit that only with high production does it count and put this into practice in their activities. [Nickum (ed.), 1981: 191.]

51. Kojima (personal discussion) cited some recent Chinese press reports to suggest that the incidence of water related disputes may have increased after 1979. In any case, they account for a substantial proportion of all legal disputes in rural areas.

52. The published collection does not include all the Chinese papers presented at the Conference. However having attended the Conference, I have access to all the papers and also made notes of proceedings. I am grateful to S. Ramanathan for preparing a summary of these papers.

REFERENCES

ADHVARYU, J.H. et al. (1980): 'Modernisation of an Irrigation System: Machu II Irrigation Project', Sardar Patel University, Vallabh Vidyanagar (mimeo.).

—— (1983): 'Socio-Economic Evaluation of an Irrigation Project: Dantiwada Project', Sardar Patel University, Vallabh Vidyanagar (mimeo.).

ADHVARYU, H.H. and A.S. Patel (1984): 'Socio-Economic Surveys of Medium Irrigation Project Areas in Gujarat: A Consolidated View', Sardar Patel University, Vallabh Vidyanagar (mimeo.).

AGARWAL, ANIL and NARAIN SUNITA, (1989): 'Towards Green Villages: A Study for Environmentally Sound and Participatory Rural Development', Centre for Science and Environment, New Delhi.

Agricultural Finance Corporation (1988): 'Report on Evaluation Study of Soil Conservation in the River Valley Project of Nizamsagar', AFC, Bombay.

Anon. 'Ralegaon Siddhi: An Experience in Watershed Development' (mimeo.).

Anon. (1958): *China's Big Leap in Water Conservancy*, Foreign Language Press, Peking.

Anon., (1996): *Theme Paper on Inter Basin Transfers of Water for National Development: Problems and Prospects*, Indian Water Resources Society, New Delhi.

ASOPA, V.N. and P.M. SHINGI (1987): 'Command Area Development Programme: Strategies for Improving Performance in India', Institute of Management, Ahmedabad (mimeo.).

BACADAYAN, ALBERT S. (1973): 'Mountain Irrigation in the Philippines', rpt. in E. Walter Coward Jr. (ed.), 1980.

BALI, J.S. (1988): 'A Critical Appraisal of Past and Present Policies and Strategies of Watershed Development and Management in India and Role of Government and Non-government Organisations in Small-scale Watershed Development', Society for Promotion of Wasteland Development (mimeo.).

BARDHAN, PRANAB K. (1993): 'Analytics of Institutions of Informal Cooperation in Rural Development', *World Development,* 21 (4).

BEARDSLEY, RICHARD K. et al. (1959): ' Japanese Irrigation Cooperatives', rpt. in Coward Jr. (ed.), 1980.

BHALLA, G.S. and Y.K. ALAGH (1979): *Performance of Indian Agriculture—A District-wise Study,* Sterling Publishers, New Delhi.

BHALLA G.S. and D.S. TYAGI (1989): *Pattern of Indian Agricultural Development: A District Level Study,* Institute of Studies in Industrial Development, New Delhi.

BHATIA, BELA (1992): 'Level Fields and Parched Throats: The Political Economy of Groundwater in Gujarat', *EPW,* Review of Agriculture, 19–26 Dec.

CARIN, ROBERT (1963): 'Irrigation Schemes in Communist China', Union Research Institute, Hong Kong (mimeo).

CARRUTHERS, IAN and COLIN CLARK (1980): *The Economics of Irrigation,* Liverpool University Press, Liverpool.

Central Research Institute for Dryland Agriculture (n.d.): 'Status Report on Model Dryland Agriculture Watersheds 1986–87', Hyderabad (mimeo.).

Central Soil and Water Research Institute (n.d.): 'Operational Research Project on Conservation Integrated Watershed Management and Institute Progress Report', Feb. 1987, Dehra Dun (mimeo.).

Centre for Monitoring Indian Economy (CMIE) (1977, 1984): Basic *Statistics Relating to the Indian Economy,* Vol. 2, States, CMIE, Bombay.

Chambers, Robert (1988): *Managing Canal Irrigation,* Oxford & IBH, New Delhi.

—— (1977): 'Basic Concepts in Organization of Irrigation', rpt. in Coward Jr. (ed.), 1980.

Chang, Chung-li (1955): *The Chinese Gentry: Studies on Their Role in the 19th Century Chinese Society,* Univ. of Washington Press, Seattle.

CHAO, KANG (1970): *Agricultural Production in Communist China 1949–1965,* University of Wisconsin, Madison.

CHATURVEDI, M.C. (1976): *Second India Series: Water,* Macmillan, New Delhi.

CHATURVEDI, M.C. and PETER RODGERS (1985): *Water Resources System Planning,* Indian Academy of Sciences, Bangalore.

CHAUHAN, B.R. (1992): *Settlement of International and Inter-State Water Disputes in India,* Indian Law Institute, New Delhi.

CHI, CHAO TING (1936): *Key Economic Areas in Chinese History,* Paragon Reprint Corp., New York, 1963.

CHOPRA, KANCHAN, GOPAL K. KADEKODI and M.N. MURTHY (1988): 'Sukhomajri and Dhamala Watersheds in Haryana: A Participatory Approach to Management', Institute of Economic Growth, Delhi (mimeo.).

——— (1990): *Participatory Development: People and Common Property Resource*, Sage, New Delhi.

CHOPRA, KANCHAN and D.V. SUBBA RAO (1997): 'Economic Evaluation of Soil and Water Conservation Programmes in Watersheds', Institute of Economic Growth, Delhi (mimeo.).

CHOPRA, KANCHAN and SEEMA BATHLA (1997): 'Water Use in the Punjab Region: Conflicts and Framework for Resolution', Paper presented at IDPAD seminar on Managing Water Scarcity: Experience and Proposals, Ameesfort, The Netherlands.

CHU, TUNGTSU (1962): *Local Government in China under Ching*, Harvard University Press, Cambridge, Mass.

COWARD JR., E. WALTER (ed.) (1980): *Irrigation and Agricultural Development in Asia*, Cornell University Press, Ithaca, N.Y.

CROOK, DAVID and ISABEL (1962): *The First Years of Yangyi Commune*, Routledge & Kegan Paul, London.

DAKSHINAMURTHY, C., A.M. MICHAEL and SHRI MOHAN (1973): *Water Resources of India and their Utilisation in Agriculture*, Water Technology Centre, IARI, New Delhi.

DANDEKAR, V.M., D. DESHMUKH and V.R. DEUSKAR (1979): *Interim Report of the Committee to Study the Introduction of Eight Monthly Supply of Water on Irrigation Projects in Maharashtra*, Government of Maharashtra, Bombay.

DATYE, K.R. (1997): *Banking on Biomass: A New Strategy for Sustainable Prosperity Based on Renewable Energy and Dispersed Industrialisation*, Centre for Environment Education, Ahmedabad.

——— (1998a): 'Approach to Managing Degraded Land', Paper presented at National Workshop on Watershed Approach for Managing Degraded Land in India, Ministry of Rural and Watershed Development, New Delhi.

——— (1998b): 'An Overview of Technique for Solving Water Problems', CASAD, Bombay (mimeo.).

DATYE, K.R. and R.K. PATIL (1987): *Farmer Managed Irrigation Systems: Indian Experience, Bombay*, Centre for Official of Systems Analysis in Development, CASAD, Bombay.

DATYE, K.R. and SUHAS PARANJAPE (1988): *Sustainable Agriculture in Semi Arid Regions: Opportunities for Small and Marginal farms*, CASAD, Bombay.

DATYE, K.R., S. PARANJAPE, V.N. GORE and K.J. JAY (1998): 'Some Important Issues involved in Watershed Development', SOPPECOM, Pune (mimeo.).

DESHPANDE, R.S. and V. RATNA REDDY (1991): 'Watershed Development Approach in Fragile Resource Region: An Analytical Study of Maharashtra', Gokhale Institute Mimeo Series, Gokhale Institute of Economics and Politics, Pune (mimeo).

DHAWAN, B.D. (1983a): 'Productivity Impact of Irrigation in India', Institute of Economic Growth, Delhi (mimeo.).

—— (1983b): 'Sourcewise Productivity of Irrigation: A Statewise Analysis', Institute of Economic Growth, Delhi (mimeo.).

—— (1986): *Economics of Groundwater Irrigation in Hard Rock Regions with Special Reference to Maharashtra*, Agricole Publishing House, New Delhi.

—— (1988): *Irrigation in India's Agricultural Development: Productivity, Stability and Equity*, Sage, New Delhi.

—— (1989): *Studies in Irrigation and Water Management*, Commonwealth Publishers, New Delhi.

—— (1990a): 'How Reliable are Groundwater Estimates?', *Economic and Political Weekly*, 25 (20).

—— (1990b): *Studies in Minor Irrigation with Special Reference to Groundwater*, Commonwealth Publishers, New Delhi.

—— (1991): 'Role of Irrigation in Raising Intensity of Cropping', *Journal of the Indian School of Political Economy*, 3 (4).

—— (1993): *Trends and New Tendencies in India's Irrigated Agriculture*, Commonwealth Publishers, New Delhi.

—— (1995): *Groundwater Depletion Land Degradation and Irrigated Agriculture in India*, Commonwealth Publishers, New Delhi.

DHRUVANARAYANA, V.V., G. SASTRY and V.S. PATNAIK (1990): *Watershed Management*, ICAR, New Delhi.

DONNITHORNE, A. (1967): *China's Economic System*, Hurst, London.

DOWNING, THEODORE E. and McGUIRE GIBSON (eds.) (1974): *Irrigation: Impact of Society*, Tuscon, University of Arizona Press.

DRÈZE, JEAN, MEERA SAMSON and SATYAJIT SINGH (eds.) (1996): *The Dam and the Nation: Displacement and Resettlement in the Narmada Valley*, Oxford University Press, New Delhi.

ELVIN, MARK (1979): 'On Water Control and Management During the Ming and Ch'ing Periods: A Review Article', *Ching-shih wen-it*, 3(3).

FEI, HSIAO-TUNG (1939): *Peasant Life in China: A Field Study of Country Life in Yangtze Valley*, Routledge, London.

FERNANDEZ, WALTER (1989): 'Tribals, Wastelands Development and Community Organisation', *in* Singh (ed.), 1989.

FERNEA, ROBERT (1970): *Shaykh and Effendi: Changing Patterns of Authority among the El Shabana of South Iraq*, Harvard University Press, Cambridge, Mass.

FRAMJI K.K. and I.K. MAHAJAN (1969): *Irrigation and Drainage in the World*, Vol. I, International Congress for Irrigation and Drainage, New Delhi.

FUKUDA (1976): *Irrigation in the World*, Tokyo University Press, Tokyo.

GADGIL, M. and RAMACHANDRA GUHA (1992): *This Fissured Land: An Ecological History of India*, Oxford University Press, New Delhi.

GADGIL, D.R. (1948): 'Economic Effect of Irrigation (Report of a Survey of Direct and Indirect Benefits of the Godavari and Pravara Canals)', Gokhale Institute of Politics and Economics, Poona.

GEERTZ, CLIFFORD (1959): 'Form and Variation in Balinese Village Structure', *American Anthropologist*, 61.

—— (1967): 'Organisation of the Balinese Subak', *in* Coward Jr. (ed.), 1980.

GLICK, THOMAS (1970): *Irrigation and Society in Medieval Valencia*, Harvard University Press, Cambridge, Mass.

Government of Andhra Pradesh (1982): *Report of the Commission for Irrigation Utilisation*, GOAP, Hyderabad.

Government of India, Committee on Plan Projects (1959): *Interim Report on Minor Irrigation*, Mysore, GOI, New Delhi.

——, Planning Commission, Committee on Plan Projects (1964): *Report on Optimum Utilization of Irrigation Potential Lower Bhavani Project of Madras State*, GOI, New Delhi.

——, Department of Science and Technology, (n.d.): *Natural Resources Data Management System: A Computer Based Decision Support System for Micro-level Spatial Planning*, GOI, New Delhi.

——, Planning Commission (1965): *Evaluation of Major Irrigation Projects: Some Case Studies*, GOI, New Delhi.

Government of India (1970): 'Joint Indo-American Team Report on Efficient Water Use and Farm Management Study in India', New Delhi (mimeo.).

——, Ministry of Science and Technology (1989): 'Report of the Working Group on Wastelands Development Sector in the Eighth Five Year Plan', New Delhi.

——, Ministry of Irrigation and Power (1972): Report of the Irrigation Commission, GOI, New Delhi.

—— National Sample Survey (1968): Report No. 74, *Eighth Round: July 1954–April 1955, Report on Land Holdings (5),* 'Rural Sector: Some Aspects of Operational Holdings', NSSO, Cabinet Secretariat, Ministry of Planning, GOI, New Delhi.

—— (1968): *Report No. 144, Seventeenth Round: September, 1961–July, 1962,* 'Tables with Notes on Some Aspects of Land Holdings in Rural Areas', NSSO, GOI, Cabinet Secretariat, Ministry of Planning, New Delhi.

—— (1976): *Number 215, Twenty Sixth Round, July 1971–September 1972*, 'Tables on Land Holdings—All India', Vol. I, NSSO, Ministry of Planning, GOI, New Delhi.

——, Programme Evaluation Organisation (1961): 'Study of the Problems of Irrigation', Planning Commission, GOI, New Delhi.

—— (1959): *Report of Minor Irrigation: Madras State*, GOI, New Delhi.

—— (1960): *Report on Minor Irrigation: Andhra Pradesh*, GOI, New Delhi.

—— (1972): *Report of the Irrigation Commission*, 1972, Ministry of Irrigation and Power, GOI, New Delhi.

—— (1981): *Sixth Five Year Plan, 1980–1985*, Planning Commission, GOI, New Delhi.

—— (1982, 1984): *Statistical Abstract, India*, Central Statistical Organisation, GOI, New Delhi.

Government of Karnataka (1988): 'Comprehensive Land-use Management Project: A Profile', GOK, Bangalore.

Government of Maharashtra (1962): *Maharashtra State Irrigation Commission Report*, GOM, Bombay.

GRAY, R.F. (1963): *The Sanjo of Tanganyika: An Anthropological Study of an Irrigation-based Society*, OUP, London.

GREER, CHARLES (1979): *Water Management in Yellow River Basin*, University of Texas, Austin.

GUHA, RAMACHANDRA (1983): 'Forestry in British and Post British India', *Economic and Political Weekly*.

GUHAN, S. (1993): 'The Cauvery River Dispute: Towards Conciliation', *Frontline*, Madras.

GUHAN, S and JOAN MENCHER (1982): 'Iruvelipattu Revisited', Madras Institute of Development Studies, Madras (mimeo.).

GUSTAFFSON, JAN ERIK (1983): *Water Resources Development in the People's Republic of China*, Royal Inst. of Technology, Stockholm.

HABIB, IRFAN (1963): *Agrarian System of Mughal India*, Asia Publishing House, Bombay.

HAMASHIMA, ATSUTOSHI (1980): 'The Organisation of Water Control in Kianguan Delta in the Ming Period', *Acta Asiatica*, No. 38.

HANUMANTHA RAO, C.H. (1976): 'Growth of Irrigation in India: An Outline of Performance and Prospects', *in* Seminar on Role of Irrigation in the Development of India's Agriculture, *Seminar Series XIII*, Indian Society of Agricultural Economics, Bombay.

HARGREAVES, GEORGE H. (1977): *World Water for Agriculture: Climate, Precipitation Probabilities and Adequacies for Rainfed Agriculture*, Utah State University, Salt Lake City.

HARRIS, D.G. (1923): *Irrigation in India*, London.

HART, HENRY (1978): 'Anarchy, Paternalism or Collective Responsibility under the Canals', *Economic and Political Weekly*, 13–51/52.

HASHIM ALI, SYED (1980): 'Integrated Water Management Above and Below the Outlet', presented at the All India Workshop on *Warabandi* held at Administrative Staff College of India, Hyderabad.

HATATE, ISAO (1978): *Irrigation Agriculture and the Landlord in Modern Japan*, Institute of Development Economies, Tokyo.

—— (1979): *Irrigation Water Rights Disputes in Japan as seen in the Azusa River System*, UN University, Tokyo.

—— (1981): *The Establishing Process of the Ogo and Yamada Canals*, UN University, Tokyo.

HSIAO, KUNG CHUAN (1960): *Rural China: Imperial Control in the 19th Century*, University of Washington Press, Seattle.

HUNT, EVA AND ROBERT C. (1974): 'Irrigation Conflict and Politics: A Mexico Case', *in* Downing and Gibson (eds), 1974.

—— (1976): 'Canal Irrigation and Local Social Organisation', *Current Anthropology*, Vol. 17.

IMAMURA, NAROAME (1980): *Land Improvement Investment and Agricultural Enterprises in Japan—As Seen in the Azusa River System*, UN University, Tokyo.

International Rice Research Institute (IRRI) (1978): *Irrigation Policy and Management in South East Asia*, IRRI, Los Bagnos.

International Irrigation Management Institute and Agrarian Research and Training Institute (1994): 'Draft Final Report: Monitoring and Evaluation of Participatory Irrigation System Management', IIMI/ARTI, Colombo.

ISHIKAWA, SHIGERU (1981): 'China's Food and Agriculture: Performance and Prospects' (mimeo).

JANAKARAJAN, S. (1996): 'Consequences of Aquifer Over Exploitation: The Case of Prosperity and Deprivation', MIDS, Madras (mimeo).

—— (1997): 'The Survival of the Fittest: Conflict on Use of Groundwater, Some Evidence from Tamil Nadu, South India', Paper presented at IDPAD seminar on Managing Water Scarcity, Ameesfort, The Netherlands.

JANAKARAJAN, S. and A. VAIDYANATHAN (1998): 'Conditions and Characteristics of Groundwater Irrigation in Tamilnadu: A Study of the Periyar Vaigai Basin', MIDS, Chennai (mimeo.).

JAY, ROBERT Q. (1979): *Javanese Villages: Social Relations in Rural Modjokuto*, MIT Press, Cambridge, Mass.

JAYARAMAN, T.K. (1980): 'Implementation of Warabandi, A Management Approach', Paper presented to the All India Workshop on *Warabandi* held at Administrative Staff College, Hyderabad.

—— (1981): 'Farmers Organisations in Surface Irrigation Projects: Two Empirical Case Studies from Gujarat', *Economic and Political Weekly*, Review of Agriculture, Sept.

JOHNSON, S.H., D.L. VERMILLION and J.A. SAGARDY (1995): *Irrigation Management Transfer: Selected Papers from the International Conference on Irrigation Management Transfer*, Wuhan, China, Sept. 1994, IIMI Colombo and FAO, Rome.

JHA, DEVAKI (1976): *Evaluation of Benefits of Irrigation, Tribeni Canal Project*, Orient Longman, Bombay.

JOSHI, B.K. (1988): *A Socio-economic Study of the Tejpura (Bangor) Watershed Development Project*, Giri Institute of Development Studies, Lucknow.

JOSHI, P.K. and N.K. TYAGI (1991): 'Sustainability of Existing Earning System in Punjab and Haryana: Some Issues in Groundwater Use', *Indian Journal of Agricultural Economics*, 46 (3).

JOSHI, SATISH (1988): 'Integrated Community and Area Development Project. Depur, Dholpur, and Shyampur: A Case Study', Seva Mandir, Udaipur (mimeo).

KADEKODI, G.K. and KANCHAN CHOPRA (1989): 'Cyclic System of Development: A Monograph', Institute of Economic Growth, Delhi (mimeo.).

KADEKODI, GOPAL and KANCHAN CHOPRA (1995): 'Operationalising Sustainable Development: A Case Study of Palamau District in India', Institute of Economic Growth, Delhi (mimeo.).

KATHPALIA, G.N. (1980): 'Rotational System of Canal Supplies and Warabandi in India', Paper presented at All India Workshop on *Warabandi*, Administrative Staff College, Hyderabad.

KELLY, WILLIAM WRIGHT (1980): 'Water Control in an Agrarian State: Irrigation Organisation in a Japanese River Basin', University Microfilms, Ann Arbor, Michigan.

—— (1980a): 'Japanese Social Science Research on Irrigation Organisation: A Review', East Asian Papers No. 30, Cornell University, New York.

KHEPAR, S.D. and J.K. SONDHI (1992): 'Water Resources Development and Management Problems and Strategy for Irrigated Agriculture in the Punjab', *Water Resource Day Proceedings*, Vol.1, Punjab Agricultural University, Ludhiana.

KIKUCHI, MASAO and YUJIRO HAYAMI (1979): *Agricultural Growth Against a Land Resource Constraint: A Comparative History of Japan, Taiwan, Korea and the Phillppines*, Tokyo Centre for Economic Research, Tokyo.

KING, F.H. (1911): *Farmers of Forty Centuries: Or Permanent Agriculture in China, Korea and Japan,* Rodale Press Inc. Erasmus, Penna reprint edition.

KOLAVALLI, SASHI, A.H. KELSO and G. NAIK (1995): *Management of Irrigation Systems: The Case of Mehar and Phopal in Gujarat, India,* Indian Institute of Management, Ahmedabad.

KULKARNI, D.N. and S.N. LELE (1980): 'Rotational Water Supply on Maharashtra', Paper presented at All India Workshop on *Warabandi* at Administrative Staff College, Hyderabad.

LEACH, E.R. (1961): *Pul Eliya: A Village in Ceylon*, Cambridge University Press, Cambridge.

—— (1961): 'Village Irrigation in the Dryzone of Sri Lanka', rpt. in Coward Jr. (ed.), 1980.

LEVINE, GILBERT (1977): 'The Relationship of Design, Operation and Management', rpt. in Coward Jr. (ed.), 1980.

LEWIS, HENRY T. (1971): 'Irrigation Societies in the Northern Phillippines: A Comparative Study of Two Phillippine Barrios', rpt. in Coward Jr. (ed.), 1980.

LUDDEN, DAVID (1978): 'Agrarian Organisation in Tinnevelly District 800–1900 AD', University Microfilm, Michigan.

—— (1979): 'Patronage and Irrigation in Tamil Nadu', *Indian Economic and Social History Review,* 16(3).

MAASS, ARTHUR and RAYMOND, L., ANDERSON (1978): '... and the Desert Shall Rejoice: Conflict Growth and Justice in Arid Environment',* MIT Press, Cambridge, Mass.

MANOR, SHAH and J. CHAMBOULEYON (eds) (1993): *Performance Management in Farmer Managed Irrigation Systems*, Proceedings of an International Workshop, International Irrigation Management Institute, Colombo.

MEINZEN DICK, RUTH and MARK SVENDSON (eds) (1991): *Future Directions for Indian Irrigation: Research and Policy Issues,* International Food Policy Research Institute, Washington DC.

MICHAEL, A.M. (1978): *Irrigation Theory and Practice*, Vikas, New Delhi.

MINHAS, B.S., K.S. PARIKH and T.N. SRINIVASAN (1974): 'Toward the Structure of a Production Function for Wheat Yields with Dated Inputs of Water', *Water Resources Research,* 10 (3).

MINHAS, B.S., K.S. PARIKH and S.A. MARGLIN (1972): *Scheduling and Operations of the Bhakra Systems: Studies in Economic and Technical Evaluation,* Statistical Publishing Society, Calcutta.

MIZUSHIMI, TSUKASA and NARA TSUYOSHI (1982): 'Social Change in a Dry Village in South India: An Interim Report', Institute of Study of Language and Culture's of Asia and Africa, Tokyo (mimeo.).

MOENCH, MARCUS, S. LUNDQVIST and DINESH KUMAR (eds) (1993): *Proceedings of the Workshop on Water Management: India's Groundwater Challenge,* Viksat, Ahmedabad, and The Pacific Institute, San Francisco.

MOORTI, T.V. (1976): 'Impact of Different Sources of Irrigation on Input–Output Relations, Crop Pattern and Farm Practices', in *Seminar on Role of Irrigation in the Development of India's Agriculture,* Seminar Series XIII, Indian Society of Agricultural Economics, Bombay.

MORTIA, AKHIRA (1974): *Studies in the History of Water Management under Ch'ing,* Tokyo, Akishobo (in Japanese).

MOTONOSUKE, AMANO (1979): *Regional Development of Chinese Agriculture,* Tokyo (in Japanese).

MUKHERJEE, CHANDAN and A. VAIDYANATHAN (1988): 'Statewise Analysis of Agricultural Growth', *in* Narain et al. (ed.), *Recent Advances in Agricultural Statistics Research,* John Wiley, New Delhi.

MYERS, RAYMOND H. (1975): 'Economic Organisation and Corporation in Modern China', *in* Yuji Muramatsu (ed.), *The Policy and Economy of China,* Late Prof. Yuji Muramatsu Commemoration Board, Tokyo.

NADKARNI, M.V. et al. (1979): *Impact of Irrigation: Canal, Well and Tank Irrigation in Karnataka,* Himalaya Publishing House, Bombay.

NINAN, K.N. and S. LAKSHMIKANTHAMMA (1994): 'Sustainable Development: The Case of Watershed Development in India', *International Journal of Sustainable Development and World Ecology.*

NEEDHAM, JOSEPH A. (1971): *Science and Civilisation in China,* Vol. IV, Cambridge University Press, Cambridge.

NICKUM, JAMES A. (1976): 'Local Water Management in the Republic of China', *China Geographer,* No. 5, rpt. in Coward Jr. (ed.), 1980.

—— (1977): 'An Instance of Local Irrigation Management in China', *Economic and Political Weekly,* 1 Oct.

—— (1978): 'Labour Accumulation in Rural China and Its Role since the Cultural Revolution', *Cambridge Journal of Economics*.

—— (ed.) (1981): 'Water Management Organisation in the People's Republic of China', rpt. in Coward Jr. (ed.), 1980.

—— (1982): *Irrigation Management in China: A Review of the Literature*, World Bank, Washington DC.

NIJMAN, CHARLES (1993): *A Management Perspective on the Performance of the Irrigation Sub Sector*, International Irrigation Management Institute, Colombo.

NINAN, K.N. (1998): 'An Assessment of European Aided Watershed Development Projects in India from the Perspective of Poverty Reduction', Working paper, Centre for Development Research, Copenhagen.

NISHIKAWA, OSAMU (1971): 'Land Improvement and Modernisation of Rural Areas in Japan', Proceedings of the Dept. of Humanities, College of General Education, *Sensan Human Geography*, Vol. 52, University of Tokyo.

NISHIMURA, AKEO (1971): *Water Utilisation and Agricultural Productivity in China: Positive Case Studies* (1949–1964), Koyo Shabo, Kyoto.

OGURA, TAKEKAZU (ed.) (1963): *Agricultural Development in Modern Japan*, FAO, Tokyo Association.

OLIVER, HENRY (1961): *Irrigation and Climate*, Edward Arnold, London.

OLSON, MANCUR (1965): *The Logic of Collective Action*, Cambridge University Press, Cambridge.

OSTROM, ELINOR (1990): *Governing the Commons: The Evolution of Institutions for Collective Action*, Cambridge University Press, Cambridge.

PANT, NIRANJAN (1981): Some Aspects of Irrigation Administration, Naya Prakash, Calcutta.

—— (1981b): 'Utilisation of Canal Water Below Outlet in Kosi Irrigation Project', *Economic and Political Weekly*, Review of Agriculture, Sept.

—— (1982): 'Major and Medium Irrigation Projects: An Analysis of Cost of Escalation and Delay in Completion', Working Paper 41, Giri Institute of Development Studies, Lucknow (mimeo.).

—— (1984): 'Organisation, Technology and Performance of Irrigation System in Uttar Pradesh', Giri Institute of Development Studies, Lucknow (mimeo.).

PARANJAPE, SUHAS and K.J. JOY (1995): *Sustainable Technology: Making the Sardar Sarovar Project Viable: A Comprehensive Proposal to Modify the Project for Greater Equity and Ecological Sustainability*, Centre for Environment Education, Ahmedabad.

PARIHAR, S.S. et al. (1990): *Water Resources of Punjab Critical for the Future of its Agriculture*, Punjab Agricultural University, Ludhiana.

PASTERNAK, BURTON (1972): *Kinship and Community in Two Taiwanese Villages*, Stanford University Press, Stanford.

PATEL, S.M. and K.V. PATEL (1971): *Economics of Tubewell Irrigation*, Indian Institute of Management, Ahmedabad.

PATIL, R.K. and S.N. LELE (1995): 'Irrigation Management Transfer: Problems in Implementation', *in* Johnson et al. (eds), 1995.

PERKINS, DWIGHT H. (1966): *Agricultural Development in China 1368–1968*, Beacon, Boston (originally published in 1932).

POTTER, JACK M. (1971): *Thai Peasant Social Structure*, Chicago University Press, Chicago.

PRICE, BARBARA (1971): 'Pre Hispanic Irrigation Agriculture in Nuclear America', *Latin America Research Review*.

RAHEJA, S.K. and A.R. BANERJEE (1988): 'Survey Methodology for Planning Watershed Development', in Yadav (ed.), 1988.

RAJAGOPAL, A. (1991): 'Water Management in Agriculture with Special Reference to Irrigation Institutions', Ph.D. thesis, Centre for Development Studies, Trivandrum.

RAMASWAMY, R. IYER (1994): 'Indian Federalism and Water Resources', *Water Resource Development*, 10 (2).

RANADE, C.G. (1980): 'impact of Cropping Pattern on Agricultural Production', *Indian Journal of Agricultural Economics*, 25(2).

RAO, K.L. (1979): *India's Water Wealth*, Orient Longman, New Delhi.

RAO, S.K. (1971): 'Inter-Regional Variations in Agricultural Growth, 1952–53 to 1964–65: A Tentative Analysis in Relation to Irrigation', *Economic and Political Weekly*, 27(3).

RATH, N. and A.K. MITRA (1987): *Economics of Irrigation in Water Scarce Regions: A Study of Maharashtra*, Gokhale Institute of Politics and Economics, Pune.

RAY, S.K. (1992): ' Development of Irrigation and Its Impact on Patterns of Land Use, Output Growth and Employment', *Journal of the Indian School of Political Economy*, 5.

REDDY, M.S. (1992): 'Water Resources Development in the 21st Century: Primary Options for India', International Hydrological Decade, Endowment Lecture, Centre for Water Resources, Anna University, Chennai.

REIDINGER, R. (1974): 'Water Management by Administrative Procedures in an Indian Irrigation System', rpt. in Coward Jr. (ed.), 1980.

ROBERTS, MICHAEL (1967): 'Traditional Customs and Irrigation Development in Sri Lanka', rpt. in Coward Jr. (ed.), 1980.

SALETH, MARIA R. (1996): *Water Institutions in India: Economics, Law and Policy,* Commonwealth Publishers, New Delhi.

SARADARAJU, A. (1941): *Economic Conditions in the Madras Presidency 1800–1850,* Madras, University of Madras.

SASAKI, SHIRO (n.d.): 'Land Development and Improvement Projects in Japan', Agriculture, Forestry and Fisheries Productivity Conference, 1959.

SATPATHY, T. (1984): *Irrigation and Economic Development,* Ashish Publishing House, New Delhi.

SENGUPTA, NIRMAL (1980): 'Indigeneous Irrigation and Irrigation Social Organisation in South Bihar', *Indian Economic and Social History Review,* Apr.–June.

—— (1982): 'Tank Irrigation in Gangetic Bihar', A.N. Sinha Institute, Patna (mimeo.).

—— (1991): *Managing Common Property: Irrigation in India and Phillippines,* Sage, New Delhi.

—— (1993): *User Frendly Irrigation Designs,* Sage, New Delhi.

SHAH, TUSHAAR (1993): 'Groundwater Markets and Irrigation Development', *Political Economy and Practical Policy,* OUP, Bombay.

—— (1997): 'The Deepening Divide: Diverse Responses to the Challenge of Groundwater Depletion in Gujarat', Paper presented at IDPAD seminar on Managing Water Scarcity, Ameesfort, The Netherlands.

SHAH, MIHIR, D. BANERJI, A.S. VIJAYASHANKAR and P. AMBASTA (1997): *India's Drylands: Tribal Societies and Development through Environmental Regeneration,* OUP, New Delhi.

SHIMPO, MITSUMO (1978): *Three Decades in Shiwa: Economic Development and Social Change in a Japanese Farming Community,* Vancouver, University of British Colombia.

SINGH, BALJIT and SREEDHER MISRA (1965): *Benefit–Cost Analysis of the Sarda Canal System,* Asia, Bombay.

SINGH, CHATRAPATHI (1991): *Water Rights and Principles of Water Resource Management,* Indian Law Institute, New Delhi.

—— (1989): 'Forestry and the Law' in Singh (ed.), 1989.

SINGH, K.K. (1980): *Warabandi for Irrigated Agriculture in India,* Central Board of Irrigation and Power, New Delhi.

—— (ed.) (1983): *Utilisation of Canal Water: A Multi-disciplinary Perspective on Irrigation,* New Delhi.

SINGH, KATAR (1988): 'Dryland Watershed Development and Management: A Case Study in Karnataka', SPWD (mimeo.).

—— (1994): *Managing Common Pool Resources: Principles and Case Studies*, Oxford University Press, New Delhi.

SINGH, PRAMOD (ed.) (1989): *Problem of Wasteland.and Forest Ecology in India*, Ashish, New Delhi.

SIVANAPPAN, R.K. et al. (1987): 'Study of Over Exploitation and Under Exploitation of Groundwater in Tamilnadu', Tamilnadu Agricultural University, Coimbatore (mimeo.).

SIVASUBRAMNIAN, K. (1995): 'Irrigation Institutions in Two Large Multi-Village Tanks of Tamil Nadu', Ph.D. Thesis, Madras Institute of Development Studies, Chennai.

Society for Promotion of Wasteland Development (1989): 'A Study of Jawaja Project', New Delhi (mimeo.).

Society for Promotion of Wastelands Development (n.d.): Proceedings of the National Workshop on Small-Scale Watershed Development at Surajkund 30 October to 1st November 1988, New Delhi.

SONACHALAM, K.S. (1963): *Benefit-Cost Evaluation of Cauvery–Mettur Project*, Research Programmes Committee, Planning Commission, Government of India, New Delhi.

SRIVATSAVA, S.T. et al. (1997): *Government Subsidies in India*, National Institute of Public Finance and Policy (NIPFP), New Delhi.

STEWARD, JULIAN H. (ed.) (1955): *Irrigation Civilisation: A Comparative Study*, Pan American Union, Washington DC.

——, 'Initiation of a Research Trend: Wittfogel's Irrigation Hypothesis', in *Society and History: Essays in Honour of K.A. Wittfogel*, ed. G.L. Ulman, Morton, New York.

SVENDSON, MARK and ASHOK GULATI (eds) (1994): *Strategic Change in Indian Irrigation*, Indian Council of Agricultural Research, New Delhi, and Washington International Food Policy Research Institute, Washington.

TAKEUCHI, SATERU (1979): 'Country Reports on Farm Water Management on Japan', in *Farm-level Water Management in Selected Asian Countries*, Asian Productivity Organisation, Tokyo.

TAMAKI, AKIRA, (1979): *Development of Local Culture and the Irrigation System of the Azsua System*, UN University, Tokyo.

—— (1977): *The Development Theory of Irrigated Agriculture*, Institute of Developing Economics, Tokyo.

TANABE, SHIGEHARU (1973): 'Historical Development of the Canal System in the

Chao Phaya Delta' (original article in Japanese), *Tonan ajio Kenkyu,* Vol. II, 182.

TAYLOR, DONALD C. and THOMAS WICKHAM (eds) (1979): *Irrigation Policy and Management of Irrigation Systems in South-East Asia,* Asian Productivity Organization, Bangkok.

UN–ESCAP (1979): 'Proceedings of the Workshop on Efficient Use and Maintenance of Irrigation Systems at the Farm level in China', *Water Resources Bulletin,* No. 51.

UN–UNESCO, (1978): *World Water Balance and Water Resources of the Earth,* UNESCO, Paris.

UPADHYAY, J.N. (1988): 'Case Studies of Small-scale Watershed in Punjab and Haryana', New Delhi (mimeo.).

VAIDYANATHAN, A. (1984): Organisation and Management of Water in China, Institute of Developing Economies, Tokyo.

—— (1987): 'India's Agricultural Development in a Regional Perspective', R.C. Dutt Memorial Lecture.

—— (1991): 'Critical Issues in Indian Irrigation', *in* Dick Meinzen and Mark Svendson (eds), 1991, *Future Directions for Indian Irrigation: Research and Policy Issues,* International Food Policy Research Institute, Washington DC.

—— (1991b): 'Integrated Watershed Development: Some Major Issues', Founders Day Lecture, Society for Promotion of Wasteland Development, New Delhi, published in *Wasteland News.*

—— (1992): *Strategy for Development of Tank Irrigation,* Madras Institute of Development Studies, Chennai.

—— (1994a): 'Second India Revisited: Water', Madras, Madras Institute of Development Studies (mimeo.).

—— (1994b): 'Transferring Irrigation Management to Farmers', *Economic and Political Weekly,* 29 (47).

—— (1996): 'Depletion of Groundwater: Some Issues', *Indian Journal of Agricultural Economics,* 51 (1/2).

—— (1998): Tanks of South India (typescript) Madras Institute of Development Studies, Chennai.

VAIDYANATHAN, A., ASHA KRISHNAKUMAR, A. RAJAGOPAL and D. VARATHARAJAN (1994): 'Impact of Irrigation on Productivity of Land', *Journal of Indian School of Political Economy,* Oct.–Dec.

VAIDYANATHAN, A. and S. JANAKARAJAN (1989): 'Management of Irrigation and its Effect on Productivity under Different Environmental and Technical Conditions: A Study of Two Surface Irrigation Systems in Tamilnadu', MIDS, Madras (mimeo.).

VANDER MEER, CANUTE (1977): 'Changing Local Patterns in a Taiwanese Irrigation System', rpt. in Coward Jr. (ed.), 1980.

VANDER VELDE, EDWARD J., 'Local Consequences of a large Irrigation System in India', rpt. in Coward Jr. (ed.), 1980.

VERMEER, E.B. (1977): *Water Conservancy and Irrigation in China: Social, Economic and Agro-technical Aspects*, Leiden University Press, Leiden.

WADE, ROBERT (1978): 'Water Supply as an Instrument of Agricultural Policy: A Case Study', *Economic and Political Weekly*, Review of Agriculture, March.

—— (1979): 'India's Changing Strategy of Irrigation Development', rpt. in Coward Jr. (ed.), 1980.

—— (1980) 'On Substituting Management for Water in Canal Irrigation: A South Indian Case', *Economic and Political Weekly*, Review of Agriculture, Dec.

—— (1980): 'Managing the Main Systems', *Economic and Political Weekly* 15 (39).

—— (1982): 'The System and Administrative and Political Corruption in Canal Irrigation in South India', *Journal of Development Studies*, 18 (3).

WHITCOMBE, E. (1972): 'Agrarian Conditions in Northern India', University of California, Berkeley.

—— (1983): 'Irrigation', *in The Cambridge Economic History of India*, Vol. 2, c 1757–c 1970, ed. Dharma Kumar Cambridge University Press, Cambridge.

WIESNER, C.J. (1970): *Climate, Irrigation and Agriculture*, Angus Robertson, Sydney.

WITTFOGEL, KARL A. (1957): *Oriental Despotism: A Study in Total Power*, Yale University Press, New Haven.

YADAV, HRIDAI RAM (ed.) (1988): *Dimensions of Wasteland Development* Concept, New Delhi.

INDEX

Adhvaryu, J.H. 82n, 83
administered prices 33, 46
afforestation 143, 155
Agarwal, Anil 151, 160
agrarian organization, in China 177
agriculture (al),
    cooperatives, in China 177, 196
    development, irrigation and vii, 15, 58
    productivity, impact of irrigation on 77–80, 82–4
    water control institutions and 1–55
agro-climatic factors, and need for water ix, 15–7, 69
agro-climatic regions, of China 185
AICRPDA technology 145
Aka river system 49n, 54n
allocation of water, 24, 31, 47, 139, 140
    in China 217–33
    decision on 26–7
    functioning of 98, 102
    management of 32–45
    'optimum' 33
    problems of 33, 34, 48
    rules 135, 136
Anbazhagan, P. 73
Anderson, Raymond L. 2, 51n, 53n
Andhra Pradesh, water requirements in 65
Asia, tank irrigation in south 38

Bacadayan, Albert S. 52n, 53n
Bali, J.S. 155, 172n
banana crop, under irrigation 73
Banerjee, A.R. 159
Barak river, potential and utilization of 70
basin boards 126
Beardsley, Richard K. 52n, 53n
beneficiaries from projects, 9, 10, 20
    involvement in management 147
    labour contribution by, in China 119, 215, 231
Bhakra Nangal, irrigation from 14, 43, 109
    reservoir 51n
    water allocation procedures from 41, 171n
Bhatia, Bela 163
Bihar, irrigated cropped area in 78
Brahmaputra river, potential and utilization of 70
brigade level committees, in China 239n, 240n
British rule, in India, irrigation development during, 11
bureaucracy, management of irrigation by 24–5, 26, 28, 45, 46, 49n, 134
    in water conservancy management in China 192, 202, 205, 237n